# ARITHMETIC MADE SIMPLE

by

## A. P. SPERLING, Ph.D.

and

## SAMUEL D. LEVINSON, M.S.

Revised by Robert R. Belge,
Department of Electrical and Computer Engineering,
Syracuse University

MADE
SIMPLE
BOOKS

A MADE SIMPLE BOOK
**DOUBLEDAY**
NEW YORK  LONDON  TORONTO  SYDNEY  AUCKLAND

A MADE SIMPLE BOOK

PUBLISHED BY DOUBLEDAY

a division of Bantam Doubleday Dell Publishing Group, Inc.
1540 Broadway, New York, New York 10036

MADE SIMPLE and DOUBLEDAY are trademarks of Doubleday,
a division of Bantam Doubleday Dell Publishing Group, Inc.


*Library of Congress Cataloging-in-Publication Data*
Sperling, A. P. (Abraham Paul), 1912–
Arithmetic made simple.
ISBN 0-385-23938-6
Includes index.
    1. Arithmetic—1961–  .  I. Levinson, Samuel D.
II. Belge, Robert R.     III. Title.
QA107.S63   1988   513   87-24716

# CONTENTS

# CHAPTER ONE

# HOW MATHEMATICIANS SOLVE PROBLEMS

## DON'T BE AFRAID

Most of us subconsciously think that we should know everything. We also think that a facility with mathematics is a sign of superior intelligence and ignorance of mathematics is equivalent to stupidity. We, of course, know that this is ridiculous, yet these subconscious misconceptions can and often do have a debilitating effect on our ability to learn. Along with these misconceptions often comes fear that the world will find us out, fear that with failure I will have to admit my stupidity even to myself.

Let's begin by saying, "I know not!" We are engaged in this endeavor to learn. The wise man knows that he knows not. If we were not ignorant, we would know everything and there would be nothing to learn. So to learn we must be ignorant and there is no shame in that. There is no shame in failure, either. We learn from failure, from our failures we will discover our errors, and this IS learning. Shame only comes from not trying.

---

### H – I – N – T
### Number 1

When studying mathematics, it's a good idea to use a pencil and paper to write down ideas and solutions. Don't try to keep a lot of things in your head because it is easy to get confused.

---

One further misconception: that there are those who have a facility for mathematics and those who do not. This is simply not true. It is true that a facility in mathematics is a function of innate intellectual ability. But if you have the intelligence to read these words, you have the intelligence to do all the mathematics in this book.

Let's assume we have three containers, one measuring 5 cups, one measuring 3 cups, and one very large container that is unmarked. Our problem is to mark the large container with lines indicating measurements from 1 to 10 cups so it will look like this:

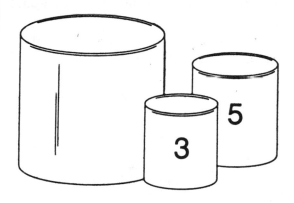

FIGURE 1.

But we only have two measurements, the 3-cup and 5-cup, so we must figure out how to use these 3- and 5-cup measures to mark all the lines on the large container.

We could fill the 3-cup measure, pour it in the large container, and draw a mark for 3 cups. Now we can fill the 3-cup measure again, pour it into the large container, and mark it. Now we have marks for 3 cups and 6 cups.

Mathematicians like to express solutions like this in an abstract way, known as "equations." For the problem just described, they would write an equation like this:

$$3 + 3 = 6$$

This is read as "3 added to 3 equals 6" or "3 plus 3 equals 6." This can also be expressed as:

$$2 \times 3 = 6$$

This is read as "2 times 3 equals 6."

Now it's your turn. Can you figure out how to make marks for 5 cups and then 10 cups using only the 3-cup and 5-cup measure? Can you write the equation for how you found the 10-cup mark? (Check your answer at the end of this chapter under A.) Now find a way to make a mark on the large container for 8 cups. Write an equation for this also. (Check your answer under B at the end of the chapter.)

To make a mark for 2 cups involves a slightly different approach. First, we would fill the 5-cup container, then pour the water into the 3-cup container until it's full. That which is left in the 5-cup container is 2 cups! We would then pour this 2-cup amount into the large container and mark it. The equation for this is:

$$5 - 3 = 2$$

This is read as "5 minus 3 equals 2."

Using this 2-cup mark and one of our original 3-cup or 5-cup measures, how would we get 7 cups marked off on the large container? Write the equation. (Check it under C.) Our large container now looks like this:

FIGURE 2.

Now do you think you can figure out how to get a mark for 1 cup? Don't be afraid, try it! (Check your answer under D.) Remember, our problem was to mark off the large container in 1-cup intervals. Do you see the general principle here? Once we have a way to make a mark for 1 cup, we can make all the other marks too. How? Just add 1 cup to 1 cup and we have two, add 1 cup to 2 and we have 3, and so on.

All right! Let's change the problem a little. We again want to mark off a large container in 1-cup intervals as before, but now instead of the 5-cup and 3-cup measure, we have a 2-cup and a 6-cup container.

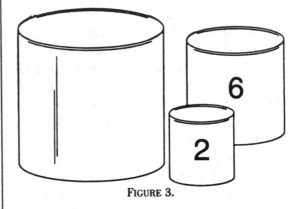

FIGURE 3.

What would be the "equation" for a 4-cup mark? How about a 10-cup mark? That's correct, $2 + 2 + 6 = 10$. Notice our equation here has three numbers to the left of the "=" sign. We could have many numbers strung together by "+" signs, indicating that we should simply add them all together. For example, we could get the 8-cup mark by $2 + 2 + 2 + 2 = 8$.

Now, another way to get the 10-cup mark would be to fill the 6-cup measure, pour it into the large container, then fill the 6-cup container again but pour enough out to fill the 2-cup container. We would then have 4 cups left in the 6-cup measure, which we would add to the large container, giving us 10 cups in the large container. Can you figure out the equation that would describe this? (Look under E at the end of the chapter.)

Remember, now, we are trying to find a way to get all the marks from 1 to 10 on the large container. This means we must get a 1-cup mark. We know that if we can get a 1-cup mark we can get all the marks. BUT now we are working with a 2-cup and 6-cup measure instead of the 3-cup and 5-cup measure.

Let's fill the 6-cup measure and pour off into the 2-cup measure. Write the equation for this. What if we now empty the 2-cup measure and again pour some of the remaining liquid from the 6-cup measure into the 2-cup measure until it's full. The equation for this is:

$$6 - (2 \times 2) = 2$$

*The parentheses mean we multiply before we subtract.* Do you see that no matter how we fill and empty from one container to the other, we cannot make the sum or difference between them come out to 1 cup?

Something different is happening here. When we had the 3-cup and the 5-cup measure, we were able to find the 1-cup mark without too much difficulty. Also recall that the key to getting all the marks was to get the 1-cup mark. With both known containers being of odd numbers (3 and 5), we could find a way to get the 1-cup mark. But now we have measures of even numbers (2 and 6), and no matter how many different ways we try, we still can't get the 1-cup mark. The numbers 1, 3, 5, 7, 9, 11, and so on are called odd numbers; the numbers 2, 4, 6, 8, 10, and so on are called even numbers. Even numbers always differ by at least 2. So we cannot combine them by addition or subtraction to get an odd number and the number 1 is an odd number.

What if the two original measures are both odd but different from 3 and 5? What if one is odd—say, 3—and the other is even—say, 8—could we then get a 1-cup mark? With the 3 and 8 we could fill the 8, then pour from the 8 into the 3, dump the 3, and then fill the 3 again from what's left

in the 8. Now the 8-cup measure has only 2 cups in it.

$$8 - (2 \times 3) = 2$$

Put the 2 into the large container and mark it; now with the 2- and 3-cup measure we can get the 1-cup. Suppose one of the original two measures was 6 and the other 3? Try this on your own. (The answer is at the end under F.) Now try to guess a general principle about what the relationship between the numbers must be to be able to get a 1-cup mark.

These are the kinds of questions mathematicians ask themselves; this is what we call "logical thinking." It's just the kind of thinking people do every day. It's the kind of thinking YOU have been doing. We think of a problem, then guess and wonder what it all means. What can we conclude from it? How does it apply to the simple case? How does it apply to the more complicated case? How does it apply to the general case? We guess, we wonder, and we guess some more; this is how we make discoveries.

The only difference between you and the so-called mathematician is how many basic arithmetic and mathematical fundamentals they have learned. So all you need is to begin learning the fundamentals and *this book will help you do just that*.

Here are the answers to the practice questions in this chapter.

**A.** $5 + 5 = 10$

**B.** $5 + 3 = 8$

**C.** $2 + 5 = 7$

**D.** Fill the 3-cup container—pour it into the large container until water is just up to the 2-cup mark—empty large container—pour the remainder of the 3-cup container into the large container and mark it. $3 - 2 = 1$

**E.** $6 + 6 - 2 = 10$

**F.** The problem with numbers like 3 and 6, even though one is odd and the other is even, is that one is a multiple of the other. No matter how we add and subtract integral multiples of these two numbers, we will always get an even number and we need an odd number, 1.

In the later chapters, practice exercises will follow each chapter. The answers to those exercises can be found beginning on page 154.

---

**H – I – N – T**
**Number 2**

When studying a subject like mathematics, it is wise not to look too far ahead in the book. You need to learn these subjects step by step, and material that you are not ready for can be frightening and discouraging. You will understand the later material as you get to it.

# CHAPTER TWO

# LEARNING TO USE OUR NUMBER SYSTEM

## HOW THEY COUNTED IN EARLY TIMES

From the very beginning of time man has been in need of a method of expressing "how many," whether it be sheep, plants, fish, etc. At first man needed only a few ways to express small quantities. But as time went on, his requirements increased and a system of numbers became essential.

Did you ever stop to wonder how the cave men indicated that they wanted or needed *one, two,* or *three* items? Judging from what we have observed among uncivilized tribes in recent times, we know that they used parts of their bodies to indicate quantities. For example, they indicated the number one by pointing to their noses, the number two by pointing to their eyes, and as time went on they learned to use their fingers to express amounts up to ten.

When primitive men wanted to describe the number of sheep in a large herd, they found it difficult to do because they lacked a number system such as we have today. Their methods were simple but intelligent, since they had no system for counting above ten. As the flock passed by they placed one stone or stick in a pile for each sheep as it passed. The number of stones or sticks on the pile then indicated the number of sheep in the flock. This was inadequate since there was no way of telling anyone else how large the flock was or for writing it on paper.

As the need for numbers increased, primitive man devised other methods of keeping records. They cut notches in wood and bone. They made scratches on the walls or on the ground in this form | | | | to substitute for the piles of stones and sticks. Others began to use dots . . . . instead of scratches. As time went on people began to use symbols similar to the dots and scratches. The Maya Indians of ancient Mexico wrote their numbers as follows:

| 1 | 2 | 3 | 4 | 5 | 6 | 7 | 8 | 9 | 10 |
|---|---|---|---|---|---|---|---|---|----|

Observe how they used the line and the dot, with the line representing five dots.

## EARLY WRITTEN NUMBERS

One of the first recorded systems for writing numbers was the use of tallies. Primitive man used his own vertical scratches and simply marked them down on the peeled bark of a tree or "papyrus" as it came to be called. Although the identity of the first group of people to use written numbers has been lost to history, we do know that the practice was begun by an ancient people living near Mesopotamia between 5000 B.C. and 4000 B.C. Their marks looked something like this: I, II, III, IIII, IIIII to represent the numbers one through five. The later Egyptians were known to have written their numbers similarly as follows:

| 1 | 2 | 3 | 4 | 5 | 6 | 7 | 8 | 9 | 0 |
|---|---|---|---|---|---|---|---|---|---|

To this day, when research workers record information which they have received from people, they use a tally system to keep their records. Of particular interest is the system we use in keeping a basketball score book. We still use the convenient tally system to count by ones and fives. But can you imagine the confusion and difficulty of trying to show the acreage of Alaska in square yards by use of the tally system.

## INTRODUCTION OF HINDU-ARABIC NUMERALS

At first, the ancients developed names for the numbers. They spoke of having *one* sheep, *two* sheep, etc. But you can see how difficult it would be to add or subtract columns of numbers expressed only in words. Thus we learn that arithmetic computation did not begin until man came to use symbols for numbers. The kinds of symbols used for numbers went through various changes starting with the simple vertical mark of ancient Mesopotamia, progressing to the combinations of the Egyptians, the familiar numerals of the Romans, and finally to our present figures.

We are indebted to the Arabs for our present method of writing numbers. For this reason, the numerals 0 through 9, the ingredients for any number combinations we wish to write, were called Arabic numbers for a long time. But more recently historians have discovered that the system of writing numbers now used by civilized people throughout the world was originated by the Hindus in India. The Arabs learned the system from the Hindus and are credited with having brought it to Europe soon after the conquest of Spain in the eighth century A.D. For this reason, we now properly call it the Hindu-Arabic system of numerals.

## READING AND WRITING ROMAN NUMERALS

An early system of writing numbers is the Roman system. It is generally agreed that it is of little practical value in today's world of advanced mathematics.

Because you will still see Roman numerals used in recording dates, in books, as numbers on a clock face, and in other places, it is worth taking a little time to learn how to read them.

The Roman number system is based on seven letters, all of which are assigned specific values. They are:

| I | V | X | L | C | D | M |
|---|---|---|---|---|---|---|
| 1 | 5 | 10 | 50 | 100 | 500 | 1000 |

Here are a few rules to help you read Roman numerals.

**Rule 1.** *When a letter is repeated, its value is repeated.*

EXAMPLES:

I = 1  II = 2  III = 3  XX = 20  CCC = 300

**Rule 2.** *When a letter follows a letter of greater value, its value is added to the greater value.*

EXAMPLES:

VI = 6  XV = 15  LX = 60  DC = 600

In these examples, observe that the smaller value I after the V means add 1 to the 5 to give 6. In the same way, the V following the X means add 5 to 10 which equals 15. Similarly, LX represents 10 added to 50 to give 60. To write 70, merely add XX after the L to give LXX. In like manner, to write 800, add CC after DC to give DCCC.

**Rule 3.** *When a letter precedes a letter of greater value, its value is subtracted from the greater value.*

EXAMPLES:

IV = 4    IX = 9    XL = 40
XC = 90    CD = 400

In these examples, note that the smaller value I, in front of the V, means subtract 1 from 5 to give 4. In the same way, the X in front of the L reduces the 50 by 10 to give

40. In like manner, X in front of C means 100 less 10 or 90 and CD denotes 500 less 100 or 400.

Generally, the symbols are not repeated more than three times to denote a number. To show the number 40 you would write XL and not XXXX. While occasionally 4 is written as IIII, it is usually written as IV.

**Rule 4.** *A horizontal bar over a letter or letters indicates that the value given to the letter or letters is to be increased one thousand times.*

EXAMPLES:

MCD = 1400    $\overline{\text{MCD}}$ = 1,400,000

Here are some additional examples of Roman numerals and their Hindu-Arabic number equivalents.

| | | |
|---|---|---|
| VII = 7 | XXII = 22 | CXIII = 113 |
| XI = 11 | XXXVII = 37 | CCX = 210 |
| XIV = 14 | XLI = 41 | $\overline{\text{MCM}}$ = 1900 |
| XVIII = 18 | LXII = 62 | $\overline{\text{XICCC}}$ = 11,300 |

You are now ready to attempt your first practice exercise. This book contains many exercises to help you determine your own rate of progress. When you complete an exercise, check your answers with those in the answers to practice exercises found on Page 154.

### Practice Exercise No. 1

Write the Roman numeral equivalents for these Hindu-Arabic numbers.

| | | | |
|---|---|---|---|
| 8 | 48 | 91 | 1958 |
| 16 | 53 | 114 | 10,200 |
| 24 | 76 | 456 | 100,000 |
| 39 | 89 | 802 | 2,000,000 |

### Practice Exercise No. 2

Write the Hindu-Arabic numbers for each of the Roman numerals below.

| | | | |
|---|---|---|---|
| VII | XLVI | LXXVIII | CV |
| XXIII | LXIX | XCII | CCXV |

| | | | |
|---|---|---|---|
| CCCLX | CMLXXI | MCMLX | $\overline{\text{CLXX}}$ |
| DCXXXI | MII | $\overline{\text{XCCC}}$ | $\overline{\text{MCMV}}$ |

## HOW "TEN" RELATES TO OUR NUMBER SYSTEM

You do not have to be a mathematician to see that it would be impractical to add, subtract, multiply, or divide using Roman numerals. For this reason they were never adopted as a basis for any system of applied mathematics.

It is generally believed that *ten* became the basis of our own number system because we have ten fingers. One evidence for this is the fact that an ancient word for finger was *digit*, and the numbers 0 through 9 have been called digits since the advent of recorded history.

## NUMBERS TO THE LEFT AND RIGHT OF THE DECIMAL POINT

With the ten digits or numerals and a decimal point, we can write numbers to represent quantities so large as to be unimaginable or so small as to be immeasurable.

It is to be noted that in the Roman numeral system there is no provision for numerical values of less than one. In our system, appropriately called a decimal system, we can indicate values of less than one by using a decimal point. The word decimal comes from the Latin word *decem*, which means ten.

The numbers which appear to the left of the decimal point are whole numbers and are called *integers*. They always have a value of one or more. For example, 6, 14, 367, or 4293 are integers.

Conversely, .6, .14, .367, .4928 are decimal fractions *placed to the right of the decimal point and have a value of less than 1*. At this time, we shall concentrate on understanding whole numbers, numbers which appear to the left of the decimal point.

## UNDERSTANDING PLACE VALUES

In our number system, you can only understand the value of a digit when you are able to recognize it in its place.

The number 9 standing alone means nine ones. With a zero (0) after it, the number becomes 90 and is read ninety. It is our way of putting the 9 two places to the left of the decimal.

The zero (0), or cipher as it is sometimes called, thus becomes a place holder.

The same 9 with two zeros after it becomes 900 and is read nine hundred. The two zeros hold down two-place values this time.

To illustrate the idea of *place values* in a different way, we can use the number 999, which we read as nine hundred ninety-nine.

In a place-value chart, we could show it this way:

| Hundreds | Tens | Ones |
| 9 | 9 | 9 |

(a) The 9 in the ones' place     =     9
(b) The 9 in the tens' place     =    90
(c) The 9 in the hundreds' place = 900
    Add them together, we get     999

From this it can be seen that as we move *to the left of the decimal point, each digit in a number is 10 times the value of the same digit immediately to its right*.

We therefore express varying quantities in our number system in two ways: (a) By the magnitude of the digit. (b) By the *place* of the digit with reference to the decimal point.

A two-digit number like 23 is two places removed from the decimal. Similarly, 542 (a three-digit number) and 6532 (a four-digit number) are three and four places removed from the decimal point. These are also referred to as two-place, three-place,

or four-place numbers, because each digit occupies a place.

Write a one-place number here _____, now a two-place number _____, now a three-place number _____, and finally a four-place number _____.

## ANALYZING NUMBERS ACCORDING TO PLACE VALUES

The number 23 is the same as saying 2 tens and 3 ones.

543 means 5 hundreds, 4 tens, 3 ones.

6532 means 6 thousands, 5 hundreds, 3 tens, 2 ones.

Try these:

58 means _____ tens and _____ ones.

734 means _____ hundreds, _____ tens, _____ ones.

9354 means _____ thousands, _____ hundreds, _____ tens, _____ ones.

It is apparent that quantities of less than 100 will be composed of one or two digits, that quantities of 100 through 999 will be composed of three digits and that quantities of 1000 through 9999 will be composed of four digits. In like manner, quantities of 10,000 through 99,999 will be composed of five digits and quantities of 100,000 through 999,999 will be composed of six digits.

By following this procedure, we can construct a table to aid in reading numbers up to the hundred billions.

## HOW TO READ LARGE NUMBERS

With a dollar sign before it, this number represents the amount of money the U.S. Government took in from all sources of income from January 1, 1985, until October 1986:

665,243,913,412

Can you read it?

**TABLE FOR READING NUMBERS**

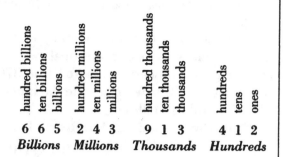

| hundred billions | ten billions | billions | hundred millions | ten millions | millions | hundred thousands | ten thousands | thousands | hundreds | tens | ones |
|---|---|---|---|---|---|---|---|---|---|---|---|
| 6 | 6 | 5 | 2 | 4 | 3 | 9 | 1 | 3 | 4 | 1 | 2 |
| *Billions* | | | *Millions* | | | *Thousands* | | | *Hundreds* | | |

We would read this number as *six hundred sixty-five billion, two hundred forty-three million, nine hundred thirteen thousand, four hundred twelve.*

Using the above table as a guide, practice reading these numbers. (The correct answers are given below.)

**1.** The total amount of money earned by the top ten tennis professionals for the first half of 1986 was 3,793,435 dollars.

**2.** Ivan Lendl was the top tennis winner for this period. He won 672,675 dollars.

**3.** The state of Alaska is estimated to cover 658,432 square miles.

**4.** A new house sold for 125,750 dollars.

**5.** The total surface area of the United States is approximately 3,623,434 square miles.

**6.** The population of the United States at the end of the year 1985 was 238,816,197.

**7.** The national deficit for the year 1985 was approximately 185,321,142,190.

To give you an idea of how big some of these numbers are, if you spend two hundred dollars a day it will take you almost fourteen years to spend one million dollars.

*Answers to above questions*
**1.** Three million, seven hundred ninety-three thousand, four hundred thirty-five.

**2.** Six hundred seventy-two thousand, six hundred seventy-five.

**3.** Six hundred fifty-eight thousand, four hundred thirty-two.

**4.** One hundred twenty-five thousand, seven hundred fifty.

**5.** Three million, six hundred twenty-three thousand, four hundred thirty-four.

**6.** Two hundred thirty-eight million, eight hundred sixteen thousand, one hundred ninety-seven.

**7.** One hundred eight-five billion, three hundred twenty-one million, one hundred forty-two thousand, one hundred ninety.

## GROUPING AND WRITING LARGE NUMBERS

When you write numbers, you will note that they are grouped in threes. In reading numbers and transcribing them to digits on paper, if you place a comma where the word billion, million, or thousand occurs, the digits will be properly grouped as you write them. Note that the comma is not used until there are five or more digits in a number.

You would write four thousand, two hundred twenty-one this way: 4221.

How would you write three hundred fifty-one? _____

### Practice Exercise No. 3

Use digits to write the indicated quantities, placing commas where needed.
**1.** Six hundred ninety-eight.

**2.** Two thousand, four hundred sixty-five.

**3.** Three thousand, four hundred twelve.

**4.** Thirty-three thousand, six hundred.

**5.** Three hundred one thousand, four hundred sixty-five.

**6.** Four hundred sixty-two thousand, three hundred nine.

**7.** Six million, four hundred twenty-two thousand, seven hundred fifty-four.

**8.** Nine billion, two million fifty.

**9.** Six hundred four million, three hundred sixty-eight thousand, four hundred nineteen.

**10.** Twenty-one billion, four hundred.

## ROUNDING OFF WHOLE NUMBERS

For convenience in using numbers and to make it easier to remember them, we often use what is known as round numbers.

**To round off a number,** you read it or write it to the nearest ten, to the nearest hundred, to the nearest thousand, or ten thousand, etc., depending upon how large the number is and what degree of accuracy is needed.

For example, 9 rounded off becomes 10.

Rounding off 63 we would write it as 60.

In rounding off 523 it might become 520 or 500, according to the exactness required.

We can better illustrate the principle of rounding off by taking a large number, for example 1,672,372.

Rounding off 1,672,372:

To the nearest
  *ten* it would be                     1,672,370
To the nearest
  *hundred* it would be                 1,672,400
To the nearest
  *thousand* it would be                1,672,000
To the nearest
  *ten thousand* it would be            1,670,000

To the nearest
  *hundred thousand* it would be        1,700,000

You can see that in rounding to the nearest ten, we dropped the final two. In rounding to the nearest hundred, the 370 was raised to 400. We can state the procedure in a rule as follows:

**Rule for Rounding Numbers:** If the digit in the final place is less than 5, drop it when rounding to the next unit on the left and replace by "0." If the digit in the final place is 5 or more, replace it by "0" and increase the next digit on the left by 1.

### Practice Exercise No. 4

This exercise will test your ability to round off numbers.

Round to the nearest ten.
  **1.** 391     **2.** 4624     **3.** 678     **4.** 8235

Round to the nearest hundred.
  **5.** 741   **6.** 6251   **7.** 82,691   **8.** 96,348

Round to the nearest thousand, then to the nearest ten thousand.
  **9.** 26,438        **11.** 388,760
  **10.** 68,770       **12.** 5,395,113

# CHAPTER THREE

# ADDITION AND SUBTRACTION OF WHOLE NUMBERS

```
H – I – N – T
Number 3
```

The big secret in doing all mathematics is feeling very comfortable with numbers, this means adding, subtracting, multiplying, and dividing. We now begin to learn these skills.

## THE LANGUAGE OF ADDITION

A shop has 32 men on the day shift and 27 on the night shift. How many men are there in both shifts?

To get the total we add:

$$+\begin{array}{r}32 \\ 27 \\ \hline 59\end{array}\quad\begin{array}{l}\text{these are \textbf{addends}} \\[4pt] \text{this is the \textbf{sum}}\end{array}$$

32 has 3 tens and 2 ones.

27 has 2 tens and 7 ones.

We add the 2 ones and the 7 ones to get 9 ones.

We add the 3 tens and the 2 tens to get 5 tens.

The *sum* is 5 tens and 9 ones or 59.

We read this as "thirty-two plus twenty-seven *is* fifty-nine."

We can write this another way:

32 + 27 = 59. The sign "+" is read **plus** and the sign "=" is read **is** or **equals**. This is in the equation form introduced in the first chapter. The entire process is called **addition**.

## SIGHT TEST IN MENTAL ADDITION

Here are 100 addition facts you should know by sight. Practice by covering the answer lines with a sheet of paper. Your answers should be written without hesitation. After each line, slip the paper down and check your answers. Circle the examples you missed.

### *Basic One Hundred Addition Facts*

| | | | | | | | | | |
|---|---|---|---|---|---|---|---|---|---|
| 3 | 5 | 4 | 7 | 2 | 3 | 3 | 5 | 3 | 1 |
| 7 | 7 | 5 | 8 | 3 | 9 | 5 | 2 | 0 | 2 |
| 10 | 12 | 9 | 15 | 5 | 12 | 8 | 7 | 3 | 3 |
| 0 | 2 | 2 | 4 | 2 | 5 | 4 | 8 | 2 | 1 |
| 2 | 6 | 2 | 4 | 5 | 0 | 9 | 8 | 4 | 5 |
| 2 | 8 | 4 | 8 | 7 | 5 | 13 | 16 | 6 | 6 |
| 9 | 8 | 0 | 6 | 7 | 2 | 1 | 6 | 8 | 5 |
| 4 | 3 | 1 | 7 | 9 | 3 | 1 | 6 | 9 | 6 |
| 13 | 11 | 1 | 13 | 16 | 5 | 2 | 12 | 17 | 11 |
| 0 | 0 | 3 | 0 | 3 | 8 | 1 | 3 | 5 | 4 |
| 4 | 0 | 8 | 3 | 3 | 6 | 6 | 6 | 8 | 6 |
| 4 | 0 | 11 | 3 | 6 | 14 | 7 | 9 | 13 | 10 |
| 6 | 4 | 2 | 9 | 0 | 7 | 1 | 5 | 7 | 6 |
| 2 | 1 | 7 | 7 | 5 | 7 | 7 | 4 | 5 | 1 |
| 8 | 5 | 9 | 16 | 5 | 14 | 8 | 9 | 12 | 7 |
| 8 | 5 | 4 | 8 | 9 | 0 | 0 | 9 | 7 | 7 |
| 0 | 3 | 0 | 4 | 1 | 7 | 6 | 0 | 4 | 0 |
| 8 | 8 | 4 | 12 | 10 | 7 | 6 | 9 | 11 | 7 |
| 2 | 8 | 6 | 5 | 7 | 8 | 8 | 2 | 1 | 6 |
| 8 | 6 | 9 | 5 | 6 | 1 | 2 | 0 | 8 | 4 |
| 10 | 14 | 15 | 10 | 13 | 9 | 10 | 2 | 9 | 10 |
| 8 | 0 | 6 | 9 | 3 | 2 | 7 | 9 | 9 | 1 |
| 5 | 8 | 5 | 2 | 4 | 9 | 3 | 8 | 9 | 3 |
| 13 | 8 | 11 | 11 | 7 | 11 | 10 | 17 | 18 | 4 |
| 1 | 3 | 4 | 5 | 9 | 5 | 3 | 1 | 0 | 1 |
| 9 | 2 | 3 | 9 | 3 | 1 | 1 | 4 | 9 | 0 |
| 10 | 5 | 7 | 14 | 12 | 6 | 4 | 5 | 9 | 1 |
| 8 | 9 | 7 | 4 | 4 | 4 | 7 | 6 | 9 | 6 |
| 7 | 6 | 2 | 2 | 8 | 7 | 1 | 3 | 5 | 0 |
| 15 | 15 | 9 | 6 | 12 | 11 | 8 | 9 | 14 | 6 |

### Use a Card System to Perfect Your Mental Addition

When you finish the 100 examples, copy the ones which gave you trouble, using both combinations as shown below, on cards with the correct answer on the back.

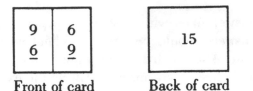

Front of card          Back of card

Below is another exercise for practice in mental addition.

### Speed Test

*Practice in Sight Addition*

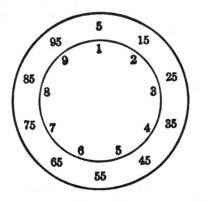

**1.** Add 1 to each figure in the outer circle; add 2 to each figure; add 3 to each figure; add 4, 5, 6, 7, 8, 9. Thus mentally you will say $1 + 5 = 6$, $1 + 15 = 16$, $1 + 25 = 26$, and so on going around the entire circle. Then add $2 + 5$, $2 + 15$, $2 + 25$, $2 + 35$, etc. Continue this until you have added every number from 1 to 9 to every number in the outer circle.

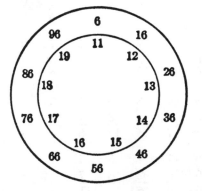

**2.** Add 11 to each figure in the outer circle. Thus mentally you will say $11 + 6 = 17$, $11 + 26 = 37$, $11 + 36 = 47$, and so on around the entire outer circle. Repeat this process for numbers from 12 through 19.

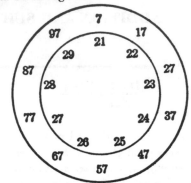

**3.** Follow same procedure as above using numbers from 21 through 29 as shown in the inner circle.

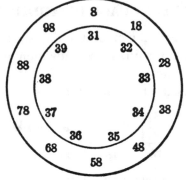

**4.** Follow same procedure as above using numbers from 31 through 39 as shown in the inner circle.

### COLUMN ADDITION THE MODERN WAY

Here is a model example in column addition.

| hundreds | tens | ones |
|:---:|:---:|:---:|
| 2 | 6 | 4 |
| 4 | 2 | 5 |
| 1 | 4 | 7 |
| 8 | 3 | 6 |

PROCEDURE: Add, from top to bottom, the numbers in the *ones' place*—think 4, 9, 16.

Since 16 is 10 ones and 6 ones, write the 6 in the ones' place of the answer. Then you mentally *exchange* the 10 ones for 1 ten and remember to add it with the numbers in the tens' column. (Some people say carry the 1 to the tens' column.) Add the numbers in the tens' column—think 7, 9, 13. Since this stands for 13 tens, write 3 in the tens' place of the answer. Again, mentally exchange the 10 of the tens for 1 hundred and remember to add it to the numbers in the hundreds' column. Finally, add the digits in the hundreds' column—think 3, 7, 8. Write the 8 in the hundreds' place of the answer.

*Check by adding up.*

## SUGGESTION AIDS FOR GOOD HABITS IN ADDING

**1.** Learn your "addition facts." Practice the list of 100 examples until you are sure of them.

**2.** Copy your numbers correctly.

**3.** Write the numbers clearly.

**4.** Keep the columns straight.

**5.** Start the addition at the right with the ones' column.

**6.** Remember to add the exchanged number to the correct column and add it *first*.

### Practice Exercise No. 5

The five problems in column addition which follow will test your skill in adding columns. For the time being you are to ignore the brackets. You will be using them later on when you learn how to add columns quickly.

| (a) | (b) | (c) | (d) | (e) |
|---|---|---|---|---|
| [48 | 58 | [42] | [39 | 76 |
| [39 | 93 | [58] | [75] | 65 |
| [56 | 58 | 37] | 93] | [48 |
| [47 | 67 | 92] | 48 | [53 |
| 85 | 77 | 74 | 91] | 79 |
| 93 | 48 | 87 | 67] | 84 |

## GAINING SPEED IN COLUMN ADDITION

One way to acquire speed in column addition is to get the habit of combining at sight two digits in the same column that make larger numbers.

Learn to pick out numbers close to each other that make 10. These are easiest to spot.

Look at the numbers in the preceding exercise. You will see, in brackets, combinations that make larger numbers such as 10, 9, 8, 7.

Try the exercise again grouping the numbers. Now do the following exercise and perform the grouping on your own.

### Practice Exercise No. 6

The exercise which follows contains five problems in column addition of large numbers. Apply what you have learned about grouping numbers.

| (a) | (b) | (c) | (d) | (e) |
|---|---|---|---|---|
| 502 | 903 | 6715 | 286 | 31,427 |
| 709 | 208 | 2564 | 2497 | 19,352 |
| 304 | 907 | 2551 | 320 | 8,911 |
| 907 | 406 | 5458 | 2493 | 70,603 |
| 106 | 900 | 3178 | 5167 | 425 |
| 703 | 705 | 2963 | 486 | 6,130 |

## HINTS ON LEARNING TO ADD MENTALLY

Let's see how many ways we can add 26 and 38.

Think 26 and 30 are 56 and 8 is 64.

Think 20 and 30 are 50; 6 and 8 are 14, 50 and 14 are 64.

Think 26 and 40 are 66; 66 less 2 is 64.

Think 20 and 38 are 58 and 6 more is 64.

### Practice Exercise No. 7

Try the following addition problems mentally, merely writing down your answers next to the equal signs. Add from left to right and check your work by adding from right to left.

1. 44 + 32 =
2. 28 + 26 =
3. 38 + 21 =
4. 22 + 29 =
5. 42 + 26 =
6. 23 + 24 =
7. 36 + 26 =
8. 24 + 36 =
9. 27 + 58 =
10. 36 + 25 =
11. 33 + 26 =
12. 28 + 38 =
13. 45 + 15 =

14. 69 + 23 =
15. 37 + 44 =
16. 24 + 23 =
17. 27 + 47 =
18. 28 + 12 =
19. 24 + 67 =
20. 93 + 19 =
21. 31 + 22 =
22. 36 + 46 =
23. 53 + 27 =
24. 28 + 44 =
25. 89 + 25 =

## COPYING NUMBERS AND ADDING

Before continuing further in your study of addition, review what you have learned so far by rereading and practicing the "suggestion aids." If you do this conscientiously, your work will improve.

### Practice Exercise No. 8

Copy the numbers into columns carefully and compute the sums. Do your work carefully and check it when you have finished.

1. 18 + 22 + 37 + 43 + 15 + 47 =
2. 84 + 36 + 15 + 27 + 62 + 48 =
3. 55 + 31 + 43 + 17 + 22 + 19 =
4. 42 + 28 + 61 + 12 + 37 + 11 =
5. 67 + 28 + 24 + 12 + 55 + 82 =
6. 268 + 149 + 438 + 324 + 646 + 423 =
7. 300 + 419 + 325 + 299 + 346 + 195 =
8. 635 + 728 + 534 + 268 + 309 + 643 + 830 =
9. 2642 + 6328 + 2060 + 9121 + 3745 =

10. 5540 + 6474 + 5567 + 2829 + 7645 =

## ADDING BY PARTIAL TOTALS

If you have long columns to add, you will find the technique of adding by use of partial totals to be most effective and accurate. When you use this method, you write down the actual sum of each column as illustrated below (do not carry any remainders to the next column) and then add the column totals to obtain the sum.

EXAMPLE:

```
          thousands
            hundreds
              tens
                ones
To Add:   6   4   5   6
          5   5   5   4
          3   5   6   6
          4   2   7   3
          8   6   2   2
          2   4   8   8
          4   2   2   9
          3   6   9   8
                  4   6   sum of ones' column
              4   4       sum of tens' column
          3   4           sum of hundreds' column
      3   5               sum of thousands' column
      3  8, 8   8   6     sum total
```

This method is helpful in checking your work. Note how simple it is to check each column.

### Practice Exercise No. 9

You are to use the partial totals method in doing the five problems which follow. Do your work carefully and check each column.

| 1 | 2 | 3 | 4 | 5 |
|---|---|---|---|---|
| 5754 | 5737 | 3594 | 2417 | 3443 |
| 2256 | 4862 | 5676 | 7989 | 5682 |
| 4445 | 6143 | 1229 | 8016 | 1317 |
| 6652 | 3688 | 8163 | 5703 | 8831 |
| 1868 | 6471 | 2223 | 4298 | 4247 |
| 6244 | 2423 | 7662 | 1683 | 4042 |
| 5471 | 1584 | 6141 | 5316 | 1761 |
| 4649 | 7845 | 8759 | 6235 | 9278 |

## THE LANGUAGE OF SUBTRACTION

PROBLEM: The baseball team started the season with 148 baseballs. At the end of a month 36 were lost. *How many were left?*

To find the answer, subtract 36 from 148.

$$
\begin{array}{rl}
148 & \text{minuend} \\
- \ 36 & \text{subtrahend} \\
\hline
112 & \text{difference}
\end{array}
$$

PROBLEM: In the second month 59 baseballs were lost. *How many more* were lost the second month than the first?

To find the answer, subtract 36 from 59.

$$
\begin{array}{rl}
59 & \text{minuend} \\
- \ 36 & \text{subtrahend} \\
\hline
23 & \text{difference}
\end{array}
$$

How many less were lost the first month than the second? _____

PROBLEM: At the end of the three-month season they had 12 baseballs left. *How many would they have to add* to start the next season with the same amount?

To find the answer, subtract 12 from 148.

$$
\begin{array}{rl}
148 & \text{minuend} \\
- \ 12 & \text{subtrahend} \\
\hline
136 & \text{difference}
\end{array}
$$

Each of these problems is solved by *subtraction*. But in each, the questions are different.

What are the questions answered by subtraction? How many ways can we say, subtract 3 from 9?

(a) How many are left when we take 3 from 9?     ANS. 6

(b) How much more is 9 than 3?     ANS. 6

(c) How much less is 3 than 9?     ANS. 6

(d) What is the difference between 3 and 9?     ANS. 6

(e) How much must be added to 3 to get 9?     ANS. 6

## SUBTRACTION VOCABULARY

Subtracting 24 from 36 leaves 12, written this way:

$$
\begin{array}{rl}
36 & \textbf{minuend,} \text{ the larger number from} \\
 & \text{which the smaller is taken} \\
- \ 24 & \textbf{subtrahend,} \text{ the number subtracted} \\
\hline
12 & \textbf{difference} \text{ or } \textbf{remainder}
\end{array}
$$

The minus sign ($-$) indicates subtraction.

### *Illustrating Subtraction to a Beginner, Using Place Values*

$$
\begin{array}{rlll}
36 = & 3 \text{ tens} & 6 \text{ ones} \\
- \ 24 = & - \ 2 \text{ tens} & 4 \text{ ones} \\
\hline
12 & 1 \text{ ten} & 2 \text{ ones} = 12
\end{array}
$$

Always begin at the right.
Take 4 ones from 6 ones leaves 2 ones.
Take 2 tens from 3 tens leaves 1 ten.
One ten and 2 ones are 12.
Try these examples with the same place-value arrangements:

$$
\begin{array}{ccccc}
48 & 34 & 56 & 75 & 89 \\
- \ 16 & - \ 21 & - \ 33 & - \ 22 & - \ 58
\end{array}
$$

## THE METHODS OF SUBTRACTION

There are two methods currently in use to solve subtraction problems. The **Exchange** or **Borrow** method is taught in most schools in the United States today and is the one which is described in detail below. The other method, the one which your parents probably learned in school, is known as the **Carry** or **Pay Back** method and is discussed briefly.

### *Subtraction Method: Exchange or Borrow*

EXAMPLE:

$$
\begin{array}{rll}
63 = & 6 \text{ tens } 3 \text{ ones} = & 5 \text{ tens } 13 \text{ ones} \\
- \ 27 = & 2 \text{ tens } 7 \text{ ones} & 2 \text{ tens } \ 7 \text{ ones} \\
\hline
 & & 3 \text{ tens } 6 \text{ ones} = 36
\end{array}
$$

$$\begin{array}{r} \overset{5\ 13}{6\ 3} \\ -\ 2\ 7 \\ \hline 3\ 6 \end{array}$$

(a) Start with the ones' place (at the right). Since we cannot subtract 7 from 3, we **exchange** one of the 6 tens for 10 ones giving 13 ones and 5 tens. Then we subtract 7 ones from 13 ones which leaves 6 ones.

(b) Next we subtract 2 tens from 5 tens, which leaves 3 tens. The **difference** is 3 tens and 6 ones or 36.

*To check your subtraction*—**add** the **difference** to the **subtrahend**. What do you get?

*Try these examples and note the exchanges.*

$$\begin{array}{ccccccc} \overset{3\ 10}{4\ 0} & \overset{7\ 13}{8\ 3} & \overset{5\ 15}{6\ 5} & \overset{4\ 13}{5\ 3} & 9\ 3 & 7\ 2 \\ -2\ 6 & -5\ 6 & -2\ 7 & -4\ 5 & -8\ 8 & -5\ 2 \end{array}$$

### ADEPTNESS IN SUBTRACTION REQUIRES DRILL AS IN ADDITION

Keep in mind the fact that subtraction is the opposite of addition. One is the *inverse* of the other.

---

NOTE:

$$\begin{array}{cccc} 9 & 15 & 6 & 15 \\ \text{If} +6 \text{ then} & -6 & \text{If} +9 \text{ then} & -9 \\ \hline 15 & 9 & 15 & 6 \end{array}$$

---

Thus if you remember your addition facts and apply them in reverse, you will know your subtraction facts. For each subtraction fact, there is a corresponding addition fact.

Here are the basic 100 subtraction facts.

### SIGHT TEST IN SUBTRACTION

Memorize these basic subtraction combinations. Practice by covering the answer. Write your answers on a blank sheet. After each line, slip the paper down and check your answers. Circle the examples on which you *hesitate* or *miss*.

### 100 Subtraction Facts

| 6 | 2 | 13 | 9 | 7 | 7 | 11 | 10 | 9 | 8 |
|---|---|---|---|---|---|---|---|---|---|
| −5 | −2 | −6 | −3 | −5 | −3 | −9 | −6 | −8 | −4 |
| 1 | 0 | 7 | 6 | 2 | 4 | 2 | 4 | 1 | 4 |

| 8 | 6 | 8 | 18 | 13 | 6 | 6 | 13 | 14 | 4 |
|---|---|---|---|---|---|---|---|---|---|
| −5 | −1 | −0 | −9 | −7 | −0 | −4 | −5 | −6 | −4 |
| 3 | 5 | 8 | 9 | 6 | 6 | 2 | 8 | 8 | 0 |

| 9 | 2 | 15 | 15 | 5 | 9 | 13 | 9 | 7 | 10 |
|---|---|---|---|---|---|---|---|---|---|
| −9 | −0 | −8 | −9 | −2 | −0 | −8 | −1 | −4 | −8 |
| 0 | 2 | 7 | 6 | 3 | 9 | 5 | 8 | 3 | 2 |

| 5 | 12 | 2 | 11 | 12 | 6 | 15 | 11 | 8 | 9 |
|---|---|---|---|---|---|---|---|---|---|
| −3 | −4 | −1 | −7 | −3 | −3 | −7 | −3 | −6 | −4 |
| 2 | 8 | 1 | 4 | 9 | 3 | 8 | 8 | 2 | 5 |

| 11 | 5 | 12 | 3 | 7 | 13 | 4 | 4 | 7 | 16 |
|---|---|---|---|---|---|---|---|---|---|
| −5 | −4 | −8 | −1 | −2 | −9 | −1 | −0 | −6 | −8 |
| 6 | 1 | 4 | 2 | 5 | 4 | 3 | 4 | 1 | 8 |

| 8 | 1 | 12 | 8 | 6 | 11 | 7 | 11 | 14 | 11 |
|---|---|---|---|---|---|---|---|---|---|
| −8 | −0 | −6 | −3 | −2 | −8 | −1 | −6 | −7 | −2 |
| 0 | 1 | 6 | 5 | 4 | 3 | 6 | 5 | 7 | 9 |

| 14 | 3 | 11 | 4 | 5 | 7 | 9 | 10 | 12 | 5 |
|---|---|---|---|---|---|---|---|---|---|
| −9 | −0 | −4 | −3 | −0 | −0 | −2 | −7 | −9 | −5 |
| 5 | 3 | 7 | 1 | 5 | 7 | 7 | 3 | 3 | 0 |

| 3 | 13 | 3 | 17 | 8 | 0 | 15 | 5 | 7 | 12 |
|---|---|---|---|---|---|---|---|---|---|
| −2 | −4 | −3 | −8 | −7 | −0 | −6 | −1 | −7 | −7 |
| 1 | 9 | 0 | 9 | 1 | 0 | 9 | 4 | 0 | 5 |

| 10 | 10 | 16 | 1 | 14 | 8 | 17 | 4 | 12 | 6 |
|---|---|---|---|---|---|---|---|---|---|
| −1 | −5 | −9 | −1 | −5 | −1 | −9 | −2 | −5 | −6 |
| 9 | 5 | 7 | 0 | 9 | 7 | 8 | 2 | 7 | 0 |

| 8 | 9 | 10 | 16 | 10 | 9 | 9 | 10 | 14 | 10 |
|---|---|---|---|---|---|---|---|---|---|
| −2 | −6 | −3 | −7 | −2 | −7 | −5 | −9 | −8 | −4 |
| 6 | 3 | 7 | 9 | 8 | 2 | 4 | 1 | 6 | 6 |

### Use Cards to Gain Speed in Subtraction

After completing the 100 subtraction examples, select those that gave you trouble and put them on study cards as you did the 100 addition facts.

The card should look like this:

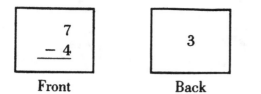

Front                    Back

With modifications, the circle arrangements which you used to practice addition may be used for additional practice in subtraction. This exercise will help to give you the needed speed and accuracy in subtraction. In each case, subtract the smaller number from the larger.

When you have become proficient in subtraction involving one and two-place numbers, you will be ready to proceed to the more difficult subtraction examples including numbers with more than two places and zeros.

## SUBTRACTION OF THREE-PLACE NUMBERS INCLUDING ZERO

$$
\begin{array}{r} 5\,6\,0 \\ -\,3\,7\,5 \end{array} =
\begin{array}{r} {}^{4\ 15\ 10}\\ 5\,6\,0 \\ -\,3\,7\,5 \\ \hline 1\,8\,5 \end{array}
$$

(a) Start with the ones' place (at the right). Since we cannot subtract 5 from 0, exchange 1 ten of the 6 tens for 10 ones. Then subtract 5 from 10, which leaves 5.

(b) Subtracting in the tens' place, 7 tens cannot be taken from 5 tens (one of the original 6 tens had been exchanged for 10 ones). Take 1 hundred of the 5 hundreds and exchange it for 10 tens, giving 15 tens. Subtract 7 tens from the 15 tens, which leaves 8.

(c) In the hundreds' place, subtract the 3 hundreds from 4 hundreds (one of the original 5 hundreds had been exchanged for 10 tens). Write the 1 in the hundreds' place. The difference is 185.

*To check your answer*—add the *difference* to the *subtrahend*. The sum should equal the *minuend. Complete the following examples and note the exchanges.*

$$
\begin{array}{ccccc}
{}^{5\,9\,11} & {}^{2\,9\,10} & & & \\
601 & 300 & 750 & 504 & 601 \\
-\,434 & -\,279 & -\,564 & -\,256 & -\,303 \\
\hline
7 & 1 & 6 & 8 & 8
\end{array}
$$

NOTE: The word "exchange" is preferred for use in teaching subtraction today. It is possible that in your school or books, the word "borrowing" or "change" is used instead. Since the word "exchange" is being adopted more and more, we will use it here.

As soon as you grasp this method of subtraction, learn to do the examples without writing the exchanges, do them mentally.

### Practice Exercise No. 10

Do the subtraction examples below without writing the exchanges. Be sure to check your work by adding the difference and the subtrahend.

|    | (a) | (b) | (c) | (d) | (e) |
|----|-----|-----|-----|-----|-----|
| 1. | 527 | 185 | 821 | 756 | 647 |
|    | − 312 | − 49 | − 337 | − 463 | − 248 |
| 2. | 147 | 536 | 289 | 343 | 426 |
|    | − 95 | − 250 | − 204 | − 59 | − 387 |
| 3. | 500 | 901 | 604 | 848 | 3005 |
|    | − 187 | − 778 | − 206 | − 792 | − 481 |
| 4. | 8909 | 8600 | 7561 | 11055 | 42211 |
|    | − 4499 | − 3075 | − 5360 | − 8037 | − 4229 |

### The Old Method of Doing Subtraction

As we said before, there is another and older method for solving subtraction problems known as the **Carry** or **Pay Back** method. Although it is not taught in many schools in the United States today, it is possible that your parents learned subtraction by this method. It works like this:

$$
\text{Subtract}\quad
\begin{array}{r}
8423 \\
-\,5445 \\
{}_{6\,5\,5}\\ \hline
2978
\end{array}
$$

The description would be as follows: You can't take 5 from 3, so borrow 1 from the 2, then 5 from 13 leaves 8. Pay back or carry the 1 to the subtrahend and 5 from 12 leaves 7, 5 from 14 leaves 9, 6 from 8 leaves 2.

This method of subtraction is only described as a matter of interest. It is *not* suggested that any modern student, taught by the exchange method, should practice this older method.

## ADDING AND SUBTRACTING WITH MONEY

We have two methods of writing money values in numerical form. We may write them using the dollar sign and decimal point—$.00 or we may write them, when the amount is less than one dollar, by using the cent sign—¢.

EXAMPLE: Eighty-three cents may be written as $.83 or 83¢.

When an amount of money consists of dollars and cents, it is always written with a dollar sign and decimal point. The decimal point separates the dollars and the cents.

EXAMPLE: One dollar and forty-three cents must be written $1.43. Three dollars can be written as $3 or $3.00.

When you add or subtract money numbers, remember to do the following:

ADD:

$$\begin{array}{r} \$13.25 \\ 6.12 \\ 3.96 \\ \underline{.15} \\ \$23.48 \end{array}$$

(a) Write in the dollar sign for the first number and the answer.

(b) Write the dollars and cents in their proper columns.

(c) There cannot be more than two numbers after the decimal point to indicate cents.

These are our money denominations.

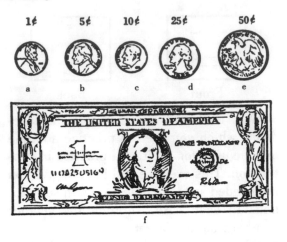

How many of each are there in a dollar?
(a) \_\_\_\_       (b) \_\_\_\_       (c) \_\_\_\_
(d) \_\_\_\_       (e) \_\_\_\_       (f) \_\_\_\_

These are the columns for money numbers:

**TABLE OF MONEY NUMBERS**

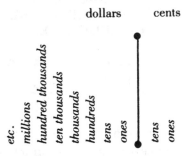

**Practice Exercise No. 11**

Using the table above as a guide, write the following money values in numerical form as dollars and cents.

1. Twelve cents

2. Six cents

3. Sixty cents

4. One hundred one cents

5. One dollar and thirty-two cents

6. Fourteen dollars and five cents

7. Two hundred twenty dollars

8. Two thousand, four hundred dollars and thirty-five cents

9. Twelve thousand, six hundred eighty-four dollars and nineteen cents

10. Three milllion, thirty dollars and ninety-eight cents

### Practice Exercise No. 12

Solve the following problems. Be sure to copy them in columns correctly.

Add and check:

1. $3.02 + $16.41 + $1.20 + $52.65 =

2. $5 + $23.64 + $16 + $.83 =

3. 37¢ + 94¢ + $4.82 + 7 cents =

4. $12.58 + $3.00 + 47¢ + $8.29 =

5. $10.32 + $15.61 + $223.14 + $6.84 + $75.38 =

Subtract and check:

6. $10.57 − $3.89 =

7. $4.50 − $2.35 =

8. $8.07 − $5.08 =

9. $19.07 − $9.38 =

10. $806.34 − $468.75 =

### Practice Exercise No. 13

The money problems which follow can all be solved by either addition or subtraction or a combination of the two. Read the problems carefully before trying to solve them.

**1.** Connie is saving to buy a U.S. Government Bond for $18.75. She has $15.30 in the bank. How much more does she need?

**2.** Cynthia does baby sitting. Last week she earned $4.25. The week before that she earned $3.50. This week she expects to earn $2.50. How much will she have earned in the three weeks?

**3.** David wants to buy swim fins for $4.75 and water goggles that cost $3.59. He has saved $3.89. How much more must he save to buy them?

**4.** Rhoda's mother bought a summer dress for $18.50 marked down from $30.00, a box of nylon stockings for $2.89 marked down from $3.00 and summer sandals at $3.50 reduced from $5.00. How much did she save by buying at the reduced prices?

**5.** The Kellys bought a new car. The advertised price with equipment was $9455.82. The dealer deducted $200 from the advertised price. They traded in their old car on which he allowed them $2475. How much did they have to pay in cash for the new car?

# CHAPTER FOUR

# MULTIPLICATION AND DIVISION OF WHOLE NUMBERS

Suppose you received four packets of Chiclets and were told there were eight Chiclets in each packet. How many Chiclets would there be in all four packets?

The quickest way to find the answer would be by multiplication, although you could also get the answer by addition.

**Multiplication** is a short method of adding a number to itself several times.

In the language of multiplication you would say "4 times 8." This means $8 + 8 + 8 + 8$ or 32. In multiplication it is written $4 \times 8 = 32$ or

$$
\begin{array}{r}
8 \\
\times\ 4 \\
\hline
32
\end{array}
\quad
\begin{array}{l}
\text{multiplicand} \\
\text{multiplier} \\
\text{product}
\end{array}
$$

The **multiplicand** is the number multiplied.

The **multiplier** indicates how many times the multiplicand is multiplied.

The **product** is the **result** which comes from multiplying one number by another.

The **sign of multiplication** is $\times$; it is read **times.**

## LEARNING MULTIPLICATION MEANS MEMORIZATION

To be proficient in multiplication you must memorize the multiplication combinations. The difficulties, in many cases, that students have in working with decimals, percentages and computing interest have been traced back to the fact that they have not memorized the multiplication combinations. The basic multiplication combinations are frequently arranged as multiplication tables to aid in memorization.

If you wish to make progress in arithmetic, you must learn to recognize by sight the product of any two numbers from 1 to 12. Below you will find the multiplication tables from 6 to 12. If you do not know them backwards and forwards, **memorize them now.** Drill them into your head before you try to go further in this book.

Work with the multiplication combinations as you did with the addition and subtraction facts. Make study cards for the combinations which give you trouble.

## MULTIPLICATION TABLE OF SIX TO TWELVE

| | | | | | | |
|---|---|---|---|---|---|---|
| $6 \times 1 = 6$ | $7 \times 1 = 7$ | $8 \times 1 = 8$ | $9 \times 1 = 9$ | $10 \times 1 = 10$ | $11 \times 1 = 11$ | $12 \times 1 = 12$ |
| $6 \times 2 = 12$ | $7 \times 2 = 14$ | $8 \times 2 = 16$ | $9 \times 2 = 18$ | $10 \times 2 = 20$ | $11 \times 2 = 22$ | $12 \times 2 = 24$ |
| $6 \times 3 = 18$ | $7 \times 3 = 21$ | $8 \times 3 = 24$ | $9 \times 3 = 27$ | $10 \times 3 = 30$ | $11 \times 3 = 33$ | $12 \times 3 = 36$ |
| $6 \times 4 = 24$ | $7 \times 4 = 28$ | $8 \times 4 = 32$ | $9 \times 4 = 36$ | $10 \times 4 = 40$ | $11 \times 4 = 44$ | $12 \times 4 = 48$ |
| $6 \times 5 = 30$ | $7 \times 5 = 35$ | $8 \times 5 = 40$ | $9 \times 5 = 45$ | $10 \times 5 = 50$ | $11 \times 5 = 55$ | $12 \times 5 = 60$ |
| $6 \times 6 = 36$ | $7 \times 6 = 42$ | $8 \times 6 = 48$ | $9 \times 6 = 54$ | $10 \times 6 = 60$ | $11 \times 6 = 66$ | $12 \times 6 = 72$ |
| $6 \times 7 = 42$ | $7 \times 7 = 49$ | $8 \times 7 = 56$ | $9 \times 7 = 63$ | $10 \times 7 = 70$ | $11 \times 7 = 77$ | $12 \times 7 = 84$ |
| $6 \times 8 = 48$ | $7 \times 8 = 56$ | $8 \times 8 = 64$ | $9 \times 8 = 72$ | $10 \times 8 = 80$ | $11 \times 8 = 88$ | $12 \times 8 = 96$ |
| $6 \times 9 = 54$ | $7 \times 9 = 63$ | $8 \times 9 = 72$ | $9 \times 9 = 81$ | $10 \times 9 = 90$ | $11 \times 9 = 99$ | $12 \times 9 = 108$ |
| $6 \times 10 = 60$ | $7 \times 10 = 70$ | $8 \times 10 = 80$ | $9 \times 10 = 90$ | $10 \times 10 = 100$ | $11 \times 10 = 110$ | $12 \times 10 = 120$ |
| $6 \times 11 = 66$ | $7 \times 11 = 77$ | $8 \times 11 = 88$ | $9 \times 11 = 99$ | $10 \times 11 = 110$ | $11 \times 11 = 121$ | $12 \times 11 = 132$ |
| $6 \times 12 = 72$ | $7 \times 12 = 84$ | $8 \times 12 = 96$ | $9 \times 12 = 108$ | $10 \times 12 = 120$ | $11 \times 12 = 132$ | $12 \times 12 = 144$ |

Write the multiplication combinations on your study cards both ways. Your cards should look like this:

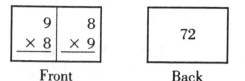

Front          Back

## MEMORY HINTS FOR THE MULTIPLICATION TABLES

(1) Note the ones' column for the products in the 8 Times Table. What similarity is there in the progress of the numbers in the ones' column as you go from $8 \times 1$ to $8 \times 5$ then from $8 \times 6$ to $8 \times 10$?

(2) Note the progression of the numbers in the ones' column for the products in the 12 Times Table. How does it compare with the ones' column of the 8 Times Table?

(3) Note the ones' column of the products in the 9 Times Table. From $9 \times 1$ to $9 \times 10$, what takes place in the ones' column?

(4) Note the ones' column of the products in the 11 Times Table. What generalization can you make?

(5) Note the 10 Times Table. To multiply any number by 10, we just add __?__ to the number.

Complete this chart. It will give you a table of multiplication combinations from 1 to 6.

|   | **0** | **1** | **2** | **3** | **4** | **5** | **6** |
|---|---|---|---|---|---|---|---|
| **0** | 0 | 0 | 0 | 0 | 0 | 0 | 0 |
| **1** | 0 | 1 | 2 | 3 | 4 | 5 | 6 |
| **2** | 0 | 2 | 4 | 6 |   |   |   |
| **3** | 0 | 3 | 6 |   |   |   |   |
| **4** | 0 | 4 |   |   |   |   |   |
| **5** | 0 | 5 |   |   |   |   |   |
| **6** | 0 | 6 |   |   |   |   |   |

From the table, any number multiplied by zero is __?__ .

Any number multiplied by 1 is __?__ .

## MULTIPLYING TWO- AND THREE-FIGURE NUMBERS

To multiply   $73 \times 3 = ?$
We could write this:

$$(7 \text{ tens} \times 3) + (3 \text{ ones} \times 3) =$$
$$21 \text{ tens} + 9 \text{ ones} =$$
$$210$$
$$\underline{+ 9}$$
$$219 \quad \text{products}$$

The parentheses mean we multiply before we add.

A shorter way of writing this:

$$\begin{array}{r} 73 \\ \times\ 3 \\ \hline 219 \end{array}$$

*Step 1.* Multiply $3 \times 3$, place the 9 in the ones' column.

*Step 2.* Multiply $3 \times 7$ tens and place the 21 tens in the tens' and hundreds' places. Product is 219.

## LEARNING "CARRYING" IN MULTIPLICATION

EXAMPLE:

$$\begin{array}{r} 63 \\ \times\ 6 \\ \hline 378 \end{array}$$

*Step 1.* Multiply $6 \times 3$ ones, which equals 18. Place the 8 under the multiplier 6 and remember the 1 ten in your mind.

*Step 2.* Multiply the $6 \times 6$ tens, which equals 36 tens. Add the 1 ten "carried over" from step 1 to get 37 tens. Write the 37 in the tens' and hundreds' columns. Product is 378.

The process is the same with a three-place number.

EXAMPLE:

$$\begin{array}{r} 353 \\ \times\ 5 \\ \hline 1765 \end{array}$$

*Step 1.* $5 \times 3$ is 15. Put down 5 and carry 1 ten.

*Step 2.* 5 × 5 is 25 + 1 is 26. Put down 6 and carry 2 hundreds.

*Step 3.* 5 × 3 is 15 + 2 is 17. Put down the 17. Product is 1765.

## MULTIPLYING WITH A ZERO IN THE MULTIPLICAND

EXAMPLE:

$$\begin{array}{r} 508 \\ \times\ 6 \\ \hline 3048 \end{array}$$

*Step 1.* 6 × 8 is 48. Write 8 in the ones column and remember to carry 4 tens.

*Step 2.* 6 × 0 is zero tens. Adding the carried over 4 tens, gives 4 in the tens place. Write it.

*Step 3.* 6 × 5 hundreds is 30. Write this in the hundreds' and thousands' columns. The product is 3048.

### Practice Exercise No. 14

Do the multiplication examples below.

| | (a) | (b) | (c) | (d) | (e) |
|---|---|---|---|---|---|
| **1.** | 43 × 3 | 20 × 4 | 32 × 7 | 66 × 8 | 24 × 4 |
| **2.** | 63 × 8 | 40 × 8 | 96 × 7 | 87 × 9 | 45 × 8 |
| **3.** | 412 × 7 | 244 × 9 | 504 × 9 | 408 × 3 | 750 × 7 |

## MULTIPLYING BY NUMBERS ENDING IN ZERO

You know that to multiply a number by 10, we place a zero to the right of the number.

EXAMPLE:

45 × 10 = 450 and 23 × 10 = 230.

**Rule: To multiply a whole number by 100,** *place two zeros to the right of the number.*

EXAMPLE:

83 × 100 = 8300 and
65 × 100 = 6500

**Rule: To multiply a whole number by 1000,** *place three zeros to the right of the number.*

EXAMPLE:

34 × 1000 = 34,000 and
56 × 1000 = 56,000

Note that 1000 has three zeros, 100 has two zeros, 10 has one zero. Therefore, what generalization can we make for multiplying by numbers ending in zero?

**Rule: To multiply by numbers ending in zero,** *multiply the numbers exclusive of the zeros, then add to the product the number of zeros at the end of the original numbers.*

EXAMPLE: 14 × 20 = ? Think
2 × 14 = 28, add 1 zero = 280

EXAMPLE: 13 × 300 = ? Think
13 × 3 = 39, add 2 zeros = 3900

Using these rules, it is possible to do many problems *mentally.*

### Practice Exercise No. 15

Do the following multiplication problems without using pencil and paper.

**1.** 40 × 5 =  
**2.** 30 × 12 =  
**3.** 40 × 50 =  
**4.** 30 × 30 =  
**5.** 50 × 30 =  
**6.** 21 × 200 =  
**7.** 20 × 80 =  
**8.** 400 × 15 =  
**9.** 31 × 300 =  
**10.** 200 × 10 =  
**11.** 8 × 3000 =  
**12.** 6000 × 20 =

## MULTIPLYING BY TWO-FIGURE NUMBERS

The proper methods for multiplying by two-figure numbers are explained on the next page. Pay close attention to both the long way and the shortcut method.

EXAMPLE:

```
 63   This is the same as:   63      63
× 24                        ×  4    × 20
 252 ← partial product ───  252    1260
1260 ← partial product ──────────────┘
1512   product
```

For greater speed, multiplication is usually done in this shorter form:

EXAMPLE:

```
   63
 × 24
  252
  126
 1512
```

*Step 1.* Multiply $63 \times 4$ as we did before. Start by writing this product in the ones' place.

*Step 2.* Multiply $63 \times 2$. Start by writing this product in the tens' place (the same column as the multiplier).

*Step 3.* Add the partial products which equal 1512.

## MULTIPLYING BY THREE-FIGURE NUMBERS

The proper methods for multiplying by three-figure numbers are explained below. You will notice that both a long way and a shortcut method are described.

EXAMPLE:

```
      708
    × 346
     4248  (a)
    28320  (b)
   212400  (c)
   244968
```

This is the same as:

```
  708        708        708
×   6      ×  40      × 300
(a) 4248   (b) 28320  (c) 212400
```

Using the short form, remember to write the right hand number in the same column as the multiplier.

EXAMPLE:

```
      708
    × 346
     4248  (a)
     2832  (b)
     2124  (c)
   244968
```

(a) You start writing the first partial product in the **ones'** place because you are multiplying by 6 **ones.**

(b) Start writing the second partial product in the **tens'** place because you are multiplying by 4 **tens.**

(c) Begin writing the third partial product in the **hundreds'** place because you are multiplying by 3 **hundreds.**

## MULTIPLYING BY A NUMBER WITH A ZERO IN THE MULTIPLIER

The correct procedures for multiplying by a number with a zero in the multiplier are explained below. Notice that both a long way and a shortcut method are described.

EXAMPLE:

```
      185
    × 106
     1110  (a)
     0000  (b)
    18500  (c)
    19610
```

This is the same as:

```
  185        185        185
×   6      ×   0      × 100
(a) 1110   (b) 000    (c) 18500
```

Using the shorter method as follows:

EXAMPLE:

```
      185
    × 106
     1110
     1850
    19610
```

NOTICE: Only *one* zero is brought down and placed in the same tens' column as the zero multiplier. The partial product for the next multiplication by 1, is placed **next** to the zero, starting in its proper **hundreds'** column. *Check* by interchanging the multiplicand and the multiplier and multiplying.

### Practice Exercise No. 16

Do the multiplication problems below.

| | | | | |
|---|---|---|---|---|
| **1.** 476 | **4.** 524 | **7.** 425 | **10.** 667 | **13.** 680 |
| × 58 | × 67 | × 143 | × 678 | × 476 |
| **2.** 6534 | **5.** 321 | **8.** 691 | **11.** 408 | **14.** 490 |
| × 47 | × 83 | × 297 | × 672 | × 508 |
| **3.** 386 | **6.** 325 | **9.** 768 | **12.** 507 | **15.** 606 |
| × 94 | × 247 | × 534 | × 305 | × 909 |

### MULTIPLICATION OF MONEY NUMBERS

When you multiply money numbers, do not forget to insert the dollar sign and decimal point in the product.

EXAMPLE:

| | |
|---|---|
| $3.85 | multiplicand |
| × 27 | multiplier |
| 2695 | partial product |
| 770 | partial product |
| $103.95 | product |

In multiplying money numbers by 10, 100, and 1000, we apply the same rules we learned before. But the method is different because of the decimal point.

Thus when multiplying money numbers by 10, move the point **one** place to the right; when multiplying by 100, move the point **two** places to the right; etc.

EXAMPLE:

$1.10 × 10 = $11.00 or $11

EXAMPLE:

$2.43 × 100 = $243.00 or $243

### Practice Exercise No. 17

Do the following multiplication examples. Since you are working with money numbers do not forget to insert the dollar signs and decimal points in the products.

| | |
|---|---|
| **1.** $3.72 × 8 | **7.** $6.50 × 20 |
| **2.** $6.24 × 19 | **8.** $4.99 × 100 |
| **3.** $7.68 × 42 | **9.** $398.23 × 100 |
| **4.** $9.35 × 76 | **10.** $561.13 × 20 |
| **5.** $2.20 × 10 | **11.** $6721.09 × 10 |
| **6.** $7.47 × 264 | **12.** $5424.06 × 100 |

### SHORTCUTS IN MULTIPLICATION

There are many shortcuts in multiplication. The rules governing multiplication when zeros are involved may be used even when the actual problem contains no zeros. Several examples of this are illustrated below.

EXAMPLE: $34 × 7 = ?$

Regroup:
$30 × 7 = 210$ and $4 × 7 = 28$
$34 × 7 = 210 + 28 = 238$ ANS.

EXAMPLE: $407 × 6 = ?$

Regroup:
$400 × 6 = 2400$ and $7 × 6 = 42$
$2400 + 42 = 2442$ ANS.

EXAMPLE: $16 × 22 = ?$

Regroup:
$16 × 20 = 320$ and $16 × 2 = 32$
$320 + 32 = 352$ ANS.

### Practice Exercise No. 18

Do the following multiplication examples mentally. Refer to the previous discussion if you have trouble.

1. $7 \times 21 =$
2. $8 \times 24 =$
3. $9 \times 37 =$
4. $6 \times 48 =$
5. $5 \times 39 =$
6. $306 \times 7 =$
7. $504 \times 8 =$
8. $408 \times 9 =$
9. $809 \times 6 =$
10. $608 \times 7 =$
11. $12 \times 24 =$
12. $15 \times 31 =$
13. $15 \times 54 =$
14. $13 \times 43 =$
15. $18 \times 32 =$

We may use our knowledge of multiplying with zeros in applying many other shortcuts in multiplication. For example:

### Multiplying Rapidly by 5, 9, or 11

EXAMPLE: $5 \times 68 = ?$

Make it $10 \times 68 = 680$. Take half and it equals 340 because 5 is half of 10.

EXAMPLE: $9 \times 17 = ?$

Make it $10 \times 17 = 170$ minus $17 = 153$. Because 9 is the equivalent of 10 minus 1.

EXAMPLE: $11 \times 17 = ?$

Make it $10 \times 17 = 170$ plus $17 = 187$. Because 11 is the equivalent of 10 plus 1.

### Multiplying by "Near" Figures:

It is often possible to multiply higher numbers more easily by working with figures that are near to even numbers.

EXAMPLE: $49 \times 26 = ?$ 49 is near 50.
Thus
$$26 \times 50 = 1300$$
$$1300 - 26 = 1274$$

EXAMPLE: $274 \times 99 = ?$
Make it
$$274 \times 100 = 27,400$$
$$27,400 - 274 = 27,126$$

### Multiplying by Numbers a Little Above or Below 100

EXAMPLE: $368 \times 106 =$
$$368 \times 100 = 36,800 \text{ and}$$
$$368 \times 6 = 2208$$
$$36,800 + 2208 = 39,008$$

EXAMPLE: $277 \times 96 =$
$$277 \times 100 = 27,700.$$
$$4 \times 277 = 1108$$
$$27,700 - 1108 = 26,592$$

### Practice Exercise No. 19

Find the products of the multiplication examples below by using the shortcut methods.

1. $67 \times 5 =$
2. $73 \times 5 =$
3. $86 \times 9 =$
4. $94 \times 9 =$
5. $78 \times 9 =$
6. $56 \times 11 =$
7. $62 \times 11 =$
8. $83 \times 11 =$
9. $34 \times 49 =$
10. $44 \times 51 =$
11. $83 \times 48 =$
12. $62 \times 52 =$
13. $68 \times 99 =$
14. $57 \times 99 =$
15. $72 \times 99 =$
16. $84 \times 101 =$
17. $71 \times 101 =$
18. $256 \times 99 =$
19. $283 \times 99 =$
20. $242 \times 101 =$
21. $149 \times 101 =$
22. $326 \times 104 =$
23. $258 \times 103 =$
24. $423 \times 97 =$
25. $352 \times 96 =$

## CHECKING RESULTS IN MULTIPLICATION

The most common method for checking a multiplication problem is to **interchange the multiplicand and multiplier** and multiply over again. But as a rule, the checking should not be more lengthy than the original problem.

EXAMPLE: You would check
$$\begin{array}{r} 473 \\ \times\ 265 \\ \hline \end{array}$$
by making it
$$\begin{array}{r} 265 \\ \times\ 473 \\ \hline \end{array}$$

But you *would not* be likely to check

$$\begin{array}{r} 48,763 \\ \times\ 23 \\ \hline \end{array}$$ by making it $$\begin{array}{r} 23 \\ \times\ 48,763 \\ \hline \end{array}$$

## DIVISION OF WHOLE NUMBERS

### A Problem in Division

There were 64 boys at the scout camp. They were to be broken up into 2 equal groups. How many would there be in each group?

To find the answer we have to divide 64 by 2. This may be written two ways:

$$2\,\overline{)\,64} \qquad \text{or} \qquad 64 \div 2$$

Written either way, the problem means that 64 is to be divided into **two equal** parts.

The quantity to be divided (64) is called the **dividend.** The number of equal parts into which it is to be divided (2) is the **divisor.** The resultant part of the division (32) is the **quotient.**

The method is as follows:

$$\begin{array}{r} 32 \\ \text{divisor} \quad 2\,\overline{)\,64} \\ \underline{6x\phantom{x}} \\ 4 \\ \underline{4} \\ 0 \end{array}$$ quotient
dividend

## HOW WE EXPLAIN DIVISION IN OUR SCHOOLS TODAY

EXAMPLE: $64 \div 2 = ?$

Regroup the 64 as follows:

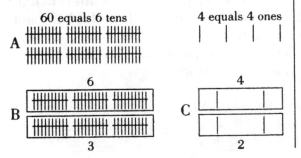

(a) We ask, how many tens and how many ones are there in 64? Answer, 6 tens and 4 ones.

(b) If we divide 6 tens into 2 equal parts, how many tens do we have in each part? Answer, 3 tens.

(c) If we divide 4 ones into 2 equal parts, how many ones do we have in each part? Answer, 2 ones.

Our quotient, therefore, is 3 tens and 2 ones or 32.

### Definitions

**Division is a process of finding equal parts of any quantity.**

Division is the reverse of multiplication. The quotient **multiplied** by the divisor will give the dividend. In the example above, $32 \times 2 = 64$.

In the problem above, division was used to find the **size** of the equal parts when the total number was given.

Division is also used to find the **number** of equal sized smaller groups contained in a larger group.

For example, suppose we had this problem. How many 6-man volleyball teams can we organize among a group of 48 boys?

METHOD: $48 \div 6 = 8$ ANS.

To check the answer, **multiply** the *quotient* (8) by the *divisor* (6). What do you get?

## WHEN THE QUOTIENT IS NOT EXACT

Sometimes the quotient is not exact. For example, consider this problem.

PROBLEM: Organize the 64 scouts into 6-man volleyball teams. How many teams would we have?

METHOD: $64 \div 6 = 10$ with 4 left over.

The number left over is called the **remainder,** and, of course, *it must be less than the divisor*. Why?

Try another: $48 \div 9 = ?$ How do we do it?

METHOD: Think, 9 times what number comes nearest to 48 and is not greater than 48? We try $9 \times 5$ and get 45. It seems all right, but you can't be sure. Try $9 \times 6$, that's 54. Too much. The answer must be 5, and since 45 is 3 less than 48, you have a remainder of 3. Thus $48 \div 9 = 5$ and 3 remainder.

### DIVISION DRILL

Facility in division requires drill in the fundamental division facts.

Since *division* is the reverse of multiplication, they may be studied together.

Turn to the multiplication tables on page 24. Read them backwards. Assume a division sign to be where the equal sign ( = ) is, and the = sign where the multiplication sign ( × ) is. Start with the 6 Times Table and read the facts from right to left (the reverse of the way you read them for multiplication purposes). Thus you would have $6 \div 1 = 6$, $12 \div 2 = 6$, $18 \div 3 = 6$, etc. Memorize them as you did the multiplication tables. Skip around. Now test yourself on the exercise below. Make study cards for those you miss.

#### Practice Exercise No. 20

Do the division examples below as rapidly as you can. This exercise will test your ability to do simple division quickly and accurately.

| | | | | | | | |
|---|---|---|---|---|---|---|---|
| 8)8 | 2)16 | 1)7 | 4)32 | 7)28 | 7)35 | 9)81 | 2)8 |
| 5)35 | 5)30 | 5)20 | 4)36 | 9)18 | 1)1 | 3)21 | 2)6 |
| 3)3 | 5)40 | 4)4 | 2)2 | 9)72 | 9)9 | 6)6 | 2)14 |
| 6)12 | 2)4 | 4)20 | 3)6 | 7)56 | 3)9 | 8)72 | 6)30 |
| 1)5 | 2)10 | 3)15 | 1)9 | 8)64 | 7)21 | 1)8 | 5)10 |
| 6)18 | 7)7 | 7)63 | 5)15 | 9)63 | 4)8 | 4)28 | 4)24 |
| 8)24 | 6)42 | 8)56 | 3)24 | 1)2 | 2)12 | 7)14 | 7)42 |
| 6)36 | 8)40 | 6)48 | 6)24 | 5)25 | 4)16 | 1)6 | 3)12 |
| 6)54 | 2)18 | 7)49 | 9)54 | 8)48 | 5)45 | 8)16 | 9)36 |
| 9)27 | 3)27 | 8)32 | 1)3 | 4)12 | 9)45 | 5)5 | 3)18 |

### STUDY PROCEDURES FOR CARRYING OUT DIVISION

There are different ways of describing the procedure in doing division, but in most cases the technique follows these steps:

EXAMPLE A: Divide 96 by 4.

*Step 1.* Estimate the final quotient in round numbers. Raise dividend to 100 and figure about 25 for quotient but less than 25, since 96 is less than 100.

$$
\begin{array}{r}
24 \\
4\overline{)96} \\
\underline{8x} \\
16 \\
\underline{16} \\
0
\end{array}
$$

*Step 2.* Think, how many times is 4 contained in 9? Try 2 as a trial quotient. Place it *over* the 9.

*Step 3.* Multiply $2 \times 4 = 8$. Subtract 8 from 9. The difference is 1.

*Step 4.* 1 is less than 4; therefore, bring down the 6.

*Step 5.* Think, how many times is 4 contained in 16? Try 4 as trial quotient. Place it over the 6.

*Step 6.* Multiply 4 in quotient by divisor. $4 \times 4 = 16$. Subtracting 16 from 16 leaves no remainder. Answer in quotient is exactly 24.

EXAMPLE B: Divide 192 by 4.

*Step 1.* Estimate. Make dividend 200 and figure quotient to be about 50 but less than 50, since 192 is less than 200.

$$
\begin{array}{r}
48 \\
4\overline{)192} \\
\underline{16x} \\
32 \\
\underline{32} \\
0
\end{array}
$$

*Step 2.* 4 is larger than 1; therefore take 19. How many times is 4 contained in 19? Try 4. Place it over the 9 of the 19.

*Step 3.* Multiply $4 \times 4 = 16$;   $19 - 16 = 3$.

*Step 4.* 3 is less than 4; therefore, bring down the 2.

*Step 5.* Think, how many times is 4 contained in 32? Try 8 as trial quotient. Place it over the 2.

*Step 6.* Multiply 8 in the quotient by divisor. $8 \times 4 = 32$. Subtracting 32 from 32 leaves no remainder. Answer in the quotient is exactly 48.

*Check* both examples above by multiplying the quotients times the divisors. Do you get the *dividend* in each case?

EXAMPLE C: Divide 327 by 4.

Estimate:

$$Try\ 4 \times 50 = 200$$
$$4 \times 80 = 320$$
$$4 \times 90 = 360$$

The answer is between 80 and 90.

$$
\begin{array}{r}
81^{R3} \\
4\overline{)327} \\
32x \\
\hline
07 \\
4 \\
\hline
3
\end{array}
$$

Proceed as in previous examples.

Three is the **remainder** because there are no additional numbers to bring down.

Write the answer as $81^{R3}$ or $81\frac{3}{4}$, which is called a **mixed number.** The $\frac{3}{4}$ part is called a **fraction.** It means 3 divided by 4. We will explain more about fractions later.

**CHECKING:** Multiply the whole-number portion of the quotient times the divisor. Then *add the remainder to the product.* Result is the dividend.

**Practice Exercise No. 21**

The exercise below will test your ability to divide by one-place numbers. Some of the examples which follow have remainders in the quotient. Work carefully and check your work.

| | | |
|---|---|---|
| **1.** $7\overline{)525}$ | **5.** $4\overline{)248}$ | **9.** $3\overline{)968}$ |
| **2.** $9\overline{)414}$ | **6.** $5\overline{)145}$ | **10.** $9\overline{)199}$ |
| **3.** $6\overline{)4926}$ | **7.** $8\overline{)2488}$ | **11.** $7\overline{)1471}$ |
| **4.** $8\overline{)4088}$ | **8.** $9\overline{)2898}$ | **12.** $6\overline{)6947}$ |

## LEARNING AIDS FOR TWO-FIGURE AND THREE-FIGURE DIVISION

To divide by two-figure or three-figure numbers, you must pay special attention to:

(a) Finding the correct partial quotient.

(b) Placement of the first-quotient figure.

(c) Use of a zero as a place holder in the quotient.

The following examples will show you how to apply these cues and carry out division with two-digit and three-digit divisors.

EXAMPLE A: Divide 736 by 32.

*Step 1.* Estimate:
$$10 \times 32 = 320$$
$$20 \times 32 = 640$$
$$30 \times 32 = 960$$

Answer is between 20 and 30, nearer to 20.

$$
\begin{array}{r}
23 \\
32\overline{)736} \\
64x \\
\hline
96 \\
96 \\
\hline
0
\end{array}
$$

*Step 2.* Divide 32 *into* 7. It can't be done. Divide 32 into 73. It can be done. To find trial quotient, think $7 \div 3$ (first number of dividend and first number of divisor) = 2. *Place this 2 over the 3 of the 73* because you are dividing 32 into 73, not 3 into 7.

*Step 3.* Multiply
$$32 \times 2 = 64;\ 73 - 64 = 9.$$

*Step 4.* 9 is less than 32, therefore bring down the 6.

*Step 5.* Think, how many times is 32 contained in 96. Divide 9 by 3 (both first digits as above) = 3. Try 3 in quotient. Place it over the 6.

*Step 6.* Multiply 3 in quotient by divisor. 3 × 32 = 96. Subtract 96 from 96 which leaves no remainder. Answer is exactly 23.

*Check* by multiplying quotient times divisor. What do you get?

EXAMPLE B: Divide 13,482 by 321.

*Step 1.* Estimate:
$$10 \times 321 = 3210$$
$$20 \times 321 = 6420$$
$$40 \times 321 = 12,840$$
$$50 \times 321 = 16,050$$

Answer is between 40 and 50, nearer to 40.

$$
\begin{array}{r}
42 \\
321 \overline{)\ 13,482} \\
1284 \\
\hline
642 \\
642 \\
\hline
0
\end{array}
$$

Follow the same procedure as above even with divisors of three or more places.
Follow this abbreviated description.

*Step 2.* 321 can't be divided into 1, 13, or 134, but can be divided into 1348. Therefore, the first number of the trial quotient goes over the 8. 13 divided by 3 = 4, try 4.

*Step 3.* 4 × 321 = 1284; 1348 − 1284 = 64.

*Step 4.* 64 is less than 321. Bring down the 2.

*Step 5.* Divide 6 by 3 (both first digits) = 2. Try 2 in quotient.

*Step 6.* 2 × 321 = 642. Subtract 642 from 642 which leaves no remainder. Answer is 42.

## HANDLING ZERO IN THE QUOTIENT

EXAMPLE: Divide 13,056 by 32

*Step 1.* Estimate:
$$200 \times 32 = 3200$$
$$400 \times 32 = 12,800$$
$$500 \times 32 = 16,000$$

Answer is between 400 and 500, nearer to 400.

$$
\begin{array}{r}
408 \\
32 \overline{)\ 13,056} \\
12\ 8 \\
\hline
256 \\
256 \\
\hline
0
\end{array}
$$

Follow the same procedure as above. Follow this abbreviated description.

*Step 2.* 32 can't be divided into 1 or 13, but it can be divided into 130. Therefore, the first number of trial quotient goes over the 0. 13 divided by 3, the result is 4, we know the answer is more than 400; so try 4.

*Step 3.* 4 × 32 = 128; 130 − 128 = 2.

*Step 4.* 2 is less than 32; bring down the 5.

*Step 5.* 25 is still less than 32; put a 0 in the quotient above the 5 and bring down the 6.

*Step 6.* 256 is bigger than 32. Divide 25 by 3, the result is a little more than 8; try 8.

*Step 7.* 8 × 32 = 256. Subtract 256 from 256, which leaves no remainder. The answer is 408.

**NOTE:** If at any step when you multiply your trial partial quotient times the divisor and the result is larger than that portion of the dividend, then the trial number is too large. Try a smaller number.

### Practice Exercise No. 22

The exercise below will test your ability to divide using two- and three-place numbers. Watch for zeros in the quotients. Check some of your answers by multiplication.

| | | | |
|---|---|---|---|
| **1.** 34 ) 8170 | **6.** 52 ) 5460 | **11.** 324 ) 8748 | **16.** 231 ) 78,540 |
| **2.** 36 ) 3492 | **7.** 46 ) 2085 | **12.** 425 ) 18,275 | **17.** 842 ) 58,940 |
| **3.** 32 ) 736 | **8.** 75 ) 3534 | **13.** 116 ) 47,098 | **18.** 180 ) 9426 |
| **4.** 64 ) 5248 | **9.** 88 ) 6450 | **14.** 235 ) 24,440 | **19.** 357 ) 20,461 |
| **5.** 24 ) 7440 | **10.** 87 ) 82,385 | **15.** 298 ) 93,572 | **20.** 581 ) 43,400 |

| No. | Sum of digits | | Remainder after ÷ 9 |
|---|---|---|---|
| 22 | 2 + 2 = | 4 | 22 ÷ 9 = 4 remainder |
| 34 | 3 + 4 = | 7 | 34 ÷ 9 = 7 remainder |
| 62 | 6 + 2 = | 8 | 62 ÷ 9 = 8 remainder |

| No. | Sum of digits | Sum of digits minus multiple of 9 | Remainder after ÷ 9 |
|---|---|---|---|
| 18 | 1 + 8 = 9 | 0 | 18 ÷ 9 = 0 remainder |
| 27 | 2 + 7 = 9 | 0 | 27 ÷ 9 = 0 remainder |
| 45 | 4 + 5 = 9 | 0 | 45 ÷ 9 = 0 remainder |
| 256 | 2 + 5 + 6 = 13 | 4 | 256 ÷ 9 = 4 remainder |
| 8645 | 8 + 6 + 4 + 5 = 23 | 5 | 8645 ÷ 9 = 5 remainder |

## CASTING OUT NINES

A quicker way of checking division than multiplying the quotient by the divisor is a method called **casting out nines.** This method is also a quick way to check multiplication. So before we continue with division, we shall learn to "cast out nines" and how to check multiplication by this method.

### Checking Multiplication by Casting Out Nines

This method of checking the accuracy of multiplication by casting out nines is based on a unique property of the number 9. That is—*the sum of the digits of a number (or the sum of these digits minus any multiple of 9) is equal to the remainder that is left after dividing the original number by nine.*

With reference to **18, 27,** and **45** (see above) note that *when the nines have been cast out of any multiple of 9 the remainder is 0.*

With reference to **256** and **8645** note that you need only add the digits in the figure representing *the sum of the original digits* to arrive at the desired remainder.

### Finding Remainders

In applying the method of casting out nines we are concerned only with *remainders*— that is, the remainder when the number is divided by 9. Using the methods indicated above, check the remainders shown here:

| | Remainder | | Remainder |
|---|---|---|---|
| 25 | 7 | 1466 | 8 |
| 35 | 8 | 16975 | 1 |
| 54 | 0 | 203468 | 5 |
| 142 | 7 | 1732159 | 1 |

### Practice Exercise No. 23

Find the remainders by casting out nines.

| | | | | | |
|---|---|---|---|---|---|
| **1.** 35 | | **3.** 126 | | **5.** 982 | |
| **2.** 87 | | **4.** 284 | | **6.** 3465 | |

| 7. 5624 | 10. 65,448 | 13. 862,425 |
| 8. 8750 | 11. 365,727 | 14. 7,629,866 |
| 9. 46,284 | 12. 584,977 | 15. 8,943,753 |

### Procedure in Checking Multiplication by Casting Out Nines

EXAMPLE:

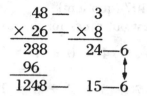

*Step 1.* Take multiplicand and cast out nines. Cast out nines in multiplier. Multiply the two remainders and cast out nines from the product. Keep *this* remainder for comparison.

*Step 2.* Cast out nines from the product of the original problem. If the *remainder* of step 2 is the same as that of step 1, the answer is probably correct.

EXAMPLE:

```
   7568            8
 × 3947          × 5
   52976         40 — 4
   30272           ↑
   68112
   22704           ↓
 29870896 — 49 — 13 — 4
```

This second example illustrates how easily the method may be applied to difficult multiplication.

The following example is purposely done incorrectly to show how a mistake is found.

Remainder should equal

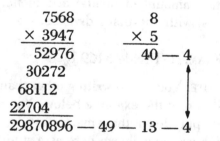

This method of checking multiplication by casting out nines is not foolproof. It can fail if the solution of a problem contains two errors that exactly offset one another. Since this type of error is not common, the method is very practical for use in checking your work.

**Practice Exercise No. 24**

Checking the accuracy of the products below by casting out nines.

1. 92 × 61 = 5612
2. 88 × 72 = 6336
3. 35 × 99 = 3464
4. 72 × 137 = 9865
5. 836 × 9321 = 7,792,356
6. 1938 × 421 = 815,893
7. 664 × 301 = 199,864
8. 736 × 428 = 315,008
9. 893 × 564 = 502,652
10. 1084 × 839 = 892,706

**Practice Exercise No. 25**

Solve the following multiplication problems.

**1.** At the theater last night all the seats were filled. We counted 65 rows with 28 people in each row. How many people were seated in the theater?

**2.** Our family bought 200 shares of stock in an auto supply corporation. The price per share was $18.75. How much did we have to pay?

**3.** Our painter gets $8.25 per hour. He estimates it will take him 56 hours. How much will it cost to paint our house?

**4.** In our backyard garden we planted 24 rows of tomatoes with 8 plants in each row. We estimate that each plant will bear 6 tomatoes. How many tomatoes in all do we estimate we will get from our planting?

**5.** In one division, a large corporation has 598 employees who receive identical salaries of $300 weekly. How much money must there be in the bank to take care of the payroll for this group for any four-week period?

### Estimating Products

For each problem below, several approximate answers are given. None is exactly correct. Select the one that is nearest the correct answer in each case.

**6.** The product of 62 times 68 is about:
(a) 6600      (b) 4200      (c) 3600      (d) 8600
(**Hint:** In estimating products, round all two-digit whole numbers to the nearest 10.)

**7.** Which is the best approximation of the product of 91 × 82?
(a) 720      (b) 7200      (c) 72,000      (d) 720,000

**8.** At a recent convention 138 organizations were represented. Each was invited to send a minimum of 50 members but not more than 60. What is the best estimate of the attendance?
(a) 8000      (b) 11,000      (c) 14,000      (d) 17,000

**9.** A National League baseball park has 1562 rows of seats. Each row seats 32 persons. The best estimate of the seating capacity is:
(a) 30,000      (b) 40,000      (c) 50,000      (d) 60,000

**10.** The sponsoring group estimated that a new automobile toll road would be used by 30,000 cars per day. They expected to collect an average toll of $1.60 from each car. What is the best estimate of the daily receipts?
(a) $10,000  (b) $30,000  (c) $50,000  (d) $70,000

### Checking Division by Casting Out Nines

Remember, we check division by multiplying the divisor by the quotient, which should give the dividend if there is no remainder. We use *"casting out nines"* to do this multiplication check.

Let's check the problem we did previously:

$$328 \overline{)15416}$$

(a) Now 328 (divisor) × 47 (quotient) should be equal to 15416 (dividend).

(b) Cast out nines for this example:

| 328 | $3 + 2 + 8 = 13$ | 4 |
|---|---|---|
| × 47 | $4 + 7 = 11$ | × 2 |
| 15416 | $1 + 5 + 4 + 1 + 6 = 17$ | 8 − —— 8 |

If there is a remainder, then a preliminary step is required. Subtract the remainder from the dividend. Now the product of the divisor and quotient should be equal to this number.

Let's check $18 \overline{)7341}$. The answer was 407 *and* R15.

(a) Subtract 15 from 7341. The new dividend 7326 should equal 18 (divisor) × 407 (quotient).

(b) Cast out nines for this example.

| 407 | $4 + 0 + 7 = 11$ | 2 |
|---|---|---|
| × 18 | $1 + 8 = 9$ | × 0 |
| 7326 | $7 + 3 + 2 + 6 = 18$ | $0 \longleftrightarrow 0$ |

For practice, check the answers to Practice Exercise No. 22 by casting out nines.

## SHORT DIVISION

Short division is a method of doing division with one- and two-place divisors by remembering most of the numbers carried forward and employing a minimum of writing.

This method is a time saver and entails a certain amount of mental arithmetic. It is easiest with one-place divisors.

EXAMPLE: Divide 3469 by 6

NOTE: You do no writing other than what is shown in the example below. You do not even put down the small carry numbers which we have shown here as a study aid.

METHOD:

$$6 \overline{)3\ 4^4 6^4 9} \quad 5\ 7\ 8^{RI}$$

DESCRIPTION: (a) Think, 6 times what is closest to 34? Answer is 5. $6 \times 5 = 30$. Put 5 in the quotient over 4 and *remember* to carry 4.

(b) Think, 6 times what is closest to 46? Answer is 7. $6 \times 7 = 42$. Put 7 in the quotient over the 6 and remember to carry 4.

(c) Think, 6 times what is closest to 49? Answer is 8. $6 \times 8 = 48$. Put 8 in the quotient over the 9 and there is a remainder of 1.

There is another way of describing the thought processes in short division. It may be familiar to some parents who were taught by this method and is worth noting here.

EXAMPLE: Divide 113,824 by 4.

$$4 \overline{)\, 1\,1^{3}3^{1}8^{2}2^{2}4} \quad \begin{array}{c} 2\,8\,4\,5\,6 \end{array}$$

DESCRIPTION: Think, 4 into 11 goes 2 times and carry 3. Write the 2 in the quotient.

Think, 4 into 33 goes 8 times and carry 1. Write 8 in the quotient.

Think, 4 into 18 goes 4 times and carry 2. Write 4 in the quotient.

Think, 4 into 22 goes 5 times and carry 2. Write 5 in the quotient.

Think, 4 into 24 goes exactly 6 times. Write 6 in the quotient. The answer is 28,456.

By now you know the multiplication and division facts of the 11 and 12 times tables. With a little additional practice you can become as proficient in multiplying and dividing by 11 and 12 as you are in multiplying and dividing by the numbers 1 through 10.

### Practice Exercise No. 26

Do the 10 problems below using the short division method.

| | |
|---|---|
| **1.** $5 \overline{)\, 3429}$ | **6.** $5 \overline{)\, 84,931}$ |
| **2.** $6 \overline{)\, 4594}$ | **7.** $8 \overline{)\, 90,412}$ |
| **3.** $7 \overline{)\, 7135}$ | **8.** $9 \overline{)\, 20,000}$ |
| **4.** $8 \overline{)\, 3653}$ | **9.** $12 \overline{)\, 25,974}$ |
| **5.** $9 \overline{)\, 6486}$ | **10.** $11 \overline{)\, 37,433}$ |

Try Practice Exercise No. 22 using the short division method. Compare your results for speed and accuracy.

## DIVISION OF MONEY NUMBERS

Do money numbers change our way of doing division? The answer is no. The only additional factor to be kept in mind is the use of the dollar sign and cents (decimal) point, which separates the dollars from the cents.

### *Placement of Point for Dollars and Cents*

When the money number is the dividend and is written with the dollar sign and cents point, write the dollar sign in the quotient and *place the cents point directly above the point in the dividend*. The correctness of the placement of the cents point can be checked because the quotient will have *only two digits* to the right of the point.

EXAMPLE: Divide $282.80 by 28.

Insert dollar sign and cents point before dividing.

$$28 \overline{)\, \$282.80} \quad \begin{array}{c} \$\quad . \end{array}$$

Now divide as usual

$$28 \overline{)\, \$282.80} \quad \begin{array}{c} \$\ 10.10 \end{array}$$
$$\underline{28x\ xx}$$
$$028$$
$$\underline{28}$$
$$00$$

Answer is $10.10.

Try another example which has a remainder.

EXAMPLE: Divide $71.24 by 53.

Insert dollar sign and cents point before dividing.

$$53 \overline{)\, \$71.24} \quad \begin{array}{c} \$\quad . \end{array}$$

Now divide as usual

$$53 \overline{)\, \$71.24} \quad \$1.34^{R22}$$

$$\begin{array}{r} \underline{53} \text{ xx} \\ 182 \\ \underline{159} \\ 234 \\ \underline{212} \\ 22 \end{array}$$

First divide 71 by 53. What does that give in the quotient?

Since 18 is smaller than 53 what is the first figure you bring down?

The next figure in the quotient goes to the right of the cents point. Why?

The answer is $1.34^{R22}$. The remainder of 22 represents $\frac{22}{53}$ of a cent.

### Practice Exercise No. 27

Do the eight problems below. Since you are working with money numbers be sure to insert the dollar signs and cents points.

1. $76 \overline{)\, \$44.08}$
2. $89 \overline{)\, \$706.60}$
3. $57 \overline{)\, \$52.44}$
4. $237 \overline{)\, \$1,395.93}$
5. $146 \overline{)\, \$386.90}$
6. $607 \overline{)\, \$4,843.86}$
7. $85 \overline{)\, \$40.80}$
8. $54 \overline{)\, \$50.22}$

### Practice Exercise No. 28

**Review Test**

The problems below will serve as an excellent review of addition, subtraction, multiplication, and division of whole numbers. Indicate which process or combination of processes should be used and then solve the problems.

1. The Borg Corp. ordered a new duplicating machine. The price including spare parts came to $2,852.67. The dealer allowed them $450 for their older equipment. How much did they have to give the dealer in cash?

2. Ellen's scout troop rented a bus to take them to camp. The cost was $416.00 which was shared equally by the 64 girls who went. How much did each girl pay?

3. A haulage truck delivered four loads of face brick to be used on a seven-story apartment building. The first load contained 3,455 usable bricks, the second 4,823, the third 3,237, and the fourth 3,684. How many usable bricks were there in these four loads?

4. The junior boys and girls of the Community Center ordered 267 sweaters with the Center insignia at a cost of $7.80 each. How much money did they have to take out of the treasury to pay for all the sweaters?

5. Henry's family started on a motor trip across the country, headed for a destination exactly 2,000 miles from their home. On Monday they drove 283 miles, Tuesday 334 miles, Wednesday 247 miles and during the next three days 970 miles. How far were they from their destination at the end of these six days?

6. When Eileen's family went on a trip during the summer vacation, they traveled for 56 days and went a distance of 10,248 miles. How many miles did they average daily?

7. Eileen's father's car goes 14 miles on a gallon of gasoline. At an average price of 89¢ per gallon, how much did her father have to pay for gasoline to travel the entire 10,248 miles?

8. The Roanoke Scout Troop undertook to wrap Red Cross packages. Each package takes 6 feet of string. They had one ball of string containing 2500 feet of string. (a) Will there be enough string for 500 packages? (b) For how many packages will there be enough string?

9. In a statewide team bowling competition, it was reported that the three leading bowlers finished with the scores of 289, 269 and 246 respectively. What was the average score for these three games?

10. Mr. Matlin, the bicycle dealer, sold 165 boy's bicycles and 157 girl's bicycles this year. The boy's bikes sold for $91.00. The girl's bikes were $8 higher. (a) How much money was taken in for the boy's bikes? (b) How much was taken in for the girl's bikes? (c) Was more or less money taken in for girl's bikes and how much more or less?

# CHAPTER FIVE

# ALL ABOUT FRACTIONS

## THE MEANING OF A FRACTION

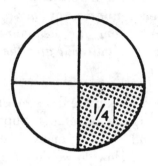

FIGURE 4.

This circle (Figure 4) is divided into four equal parts. To indicate that we are concerned with *one* of these *four* parts we write $\frac{1}{4}$. A number of this kind is called a **fraction.**

A *fraction* by definition is: *a part of any object, quantity, or digit.*

There are two numbers in a simple fraction. The *numerator* is the number on top and indicates a proportion of the whole or group. The *denominator* is the bottom number and it tells how many equal parts there are in the whole or in the group. These are the *parts* of a fraction.

In our example:

$$\text{fraction line} \rightarrow \frac{1}{4} \begin{array}{l}\text{— numerator} \\ \text{— denominator}\end{array}$$

Notice the *fraction line*. This indicates that the top number is to be divided by the bottom number.

## KINDS OF FRACTIONS

A **proper fraction** is one in which the numerator is *smaller* than the denominator.

This gives it a value of less than one, such as:

$$\frac{1}{8}, \frac{1}{4}, \frac{2}{3}, \frac{3}{4}, \frac{5}{6}.$$

An **improper fraction** is one in which the numerator is either equal to or *larger* than the denominator. This means that an improper fraction has a value of one or more than one, such as:

$$\frac{8}{8}, \frac{9}{8}, \frac{4}{3}.$$

These are improper fractions because: $(8 \div 8)$, $(9 \div 8)$, and $(4 \div 3)$ all have quotients greater than or equal to one.

An improper fraction is composed of a whole number or a whole number and a fraction.

We call it a **mixed number** when a whole number and a fraction are written together, as for example:

$$1\frac{1}{8} \text{ or } 1\frac{3}{4}$$

## USES OF FRACTIONS

**1.** Fractions are used to help us find the size or value of part of a sum of money.

EXAMPLE: John gets 9 dollars per day to take care of all his daily expenses; he may not spend more than $\frac{1}{3}$ for his lunch. To take $\frac{1}{3}$ of 9 dollars means dividing it into 3 parts. $9 \div 3 = 3$. One of those three parts or $\frac{1}{3}$ is therefore $3. ANS.

**2.** Fractions are used to help us find what part one number is of another.

EXAMPLE: George has three candy bars which are to be shared equally by four friends. What part or how much will each one get? (ANS. $\frac{1}{4}$ of 3 or $\frac{3}{4}$ of a bar for each one.)

**3.** Fractions are used to help us find values of whole quantities when we know only a part.

EXAMPLE: A half bushel of peaches costs $5.25. How much would a whole bushel cost? (ANS. $10.50.)

**4.** Fractions are used to help express facts by comparisons.

EXAMPLE: The population of one city is 60,000. We are told that a neighboring city has a population $\frac{2}{3}$ this size. What is the population of the smaller city? (ANS. 40,000.)

**5.** A fraction always means division.

EXAMPLE: $\frac{9}{5}$ means 9 divided by 5 or $5\overline{)9}$ which is equal to $1\frac{4}{5}$. We read it as "one and four-fifths." (This is the method used to change improper fractions into mixed numbers.)

## EQUIVALENT FRACTIONS

From the diagrams below (Figure 5), it can be readily seen that $\frac{1}{2}$ in A is the same as $\frac{2}{4}$ in B or $\frac{4}{8}$ in C or $\frac{3}{6}$ in D and $\frac{6}{12}$ in E.

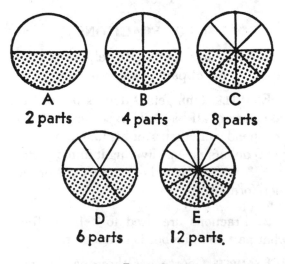

A
**2 parts**

B
**4 parts**

C
**8 parts**

D
**6 parts**

E
**12 parts**

FIGURE 5.

These fractions $\frac{1}{2}$, $\frac{2}{4}$, $\frac{4}{8}$, $\frac{3}{6}$, $\frac{6}{12}$ are all equal in value. Thus, even though the terms of the fraction $\frac{6}{12}$ are obviously greater than

the terms of the fraction $\frac{1}{2}$, the fractions are equal.

In dealing with fractions, it is important that you learn to raise fractions to higher terms or reduce them to lower terms without changing the values.

## RAISING FRACTIONS TO HIGHER TERMS

**Rule: To raise a fraction to one that has a higher denominator,** *multiply* both *the numerator and the denominator by the* same *number that will give the desired denominator.*

EXAMPLE: You can see from the diagrams above that $\frac{1}{4}$ is the same as $\frac{2}{8}$. Suppose we wanted to find out how many 24ths were equal to $\frac{1}{4}$ or $\frac{2}{8}$. How do we do it? This is a problem in raising a fraction to higher terms.

METHOD: Since 4 is now the denominator, we ask, 4 times ? equals 24. The answer is 6. Therefore we multiply by the number 6.

$$\frac{1 \times 6}{4 \times 6} = \frac{6}{24}$$

Now for $\frac{2}{8}$. Similarly, 8 times ? = 24. The answer is 3. Therefore we multiply by the number 3.

$$\frac{2 \times 3}{8 \times 3} = \frac{6}{24}$$

NOTE: *Multiplying the numerator and the denominator of any fraction by the same number does not change its value.*

### Practice Exercise No. 29

Raise the following fractions to higher terms.

**1.** $\frac{1}{3} = \frac{?}{9}$      **6.** $\frac{3}{6} = \frac{?}{36}$      **11.** $\frac{2}{11} = \frac{6}{?}$

**2.** $\frac{2}{3} = \frac{?}{12}$      **7.** $\frac{2}{7} = \frac{?}{14}$      **12.** $\frac{3}{12} = \frac{15}{?}$

**3.** $\frac{1}{4} = \frac{?}{16}$      **8.** $\frac{1}{5} = \frac{?}{25}$      **13.** $\frac{4}{9} = \frac{36}{?}$

**4.** $\frac{3}{4} = \frac{?}{20}$      **9.** $\frac{3}{8} = \frac{?}{24}$      **14.** $\frac{5}{7} = \frac{25}{?}$

**5.** $\frac{1}{6} = \frac{?}{18}$      **10.** $\frac{1}{12} = \frac{?}{48}$      **15.** $\frac{5}{6} = \frac{35}{?}$

## REDUCING FRACTIONS TO LOWER TERMS

**Rule: To reduce fractions,** *divide both the numerator and the denominator by the same number.*

The process is the opposite of raising the terms of a fraction. It is a process carried out very often in dealing with fractions because it is generally considered advisable to reduce fractions to their *lowest terms*.

EXAMPLE: Reduce $\frac{6}{24}$ to 8ths.

METHOD: Ask yourself, 24 ÷ ? equals 8. The answer is 3. Therefore

$$\frac{6}{24} \, \frac{\div \, 3}{\div \, 3} = \frac{2}{8} \text{ Ans.}$$

Suppose you wanted to reduce $\frac{6}{24}$ to 6ths. The divisor would be 4. Why?

$$\frac{6}{24} \, \frac{\div \, 4}{\div \, 4} = \frac{1\frac{1}{2}}{6}$$

This results in what we call a **complex fraction,** *one in which either the numerator or the denominator is a fraction or a mixed number.* Generally we avoid complex fractions because they are more complicated to work with.

Most often it is desirable to reduce a fraction to its *lowest terms*. This means bringing the fraction down to the smallest whole number in the numerator and the denominator without changing its value.

**Rule: To reduce to lowest terms,** *divide both the numerator and the denominator of a fraction by the* highest *number that can be divided evenly into both.*

EXAMPLE: Reduce $\frac{16}{24}$ to lowest terms.

What numbers can be divided into 16?
2, 4, $\boxed{8}$, 16

What numbers can be divided into 24?
2, 3, 4, 6, $\boxed{8}$, 12, 24

The largest *even divisor* of both 16 and 24 is 8.

$$\frac{16}{24} \, \frac{\div \, 8}{\div \, 8} = \frac{2}{3}. \quad \text{Thus } \frac{16}{24} = \frac{2}{3} \text{ Ans.}$$

NOTE: *Dividing the numerator and the denominator of any fraction by the same number does not change its value.*

The test to determine whether a fraction is reduced to its lowest terms is to see whether there is any number that can be divided evenly into both numerator and denominator. For example, try the $\frac{16}{24}$ from above and suppose we stopped at 4, thinking it was the largest divisor. In this way

$$\frac{16}{24} \, \frac{\div \, 4}{\div \, 4} = \frac{4}{6}$$

From this you can see that an additional division by 2 would be needed to bring it down to lowest terms.

$$\frac{4}{6} \, \frac{\div \, 2}{\div \, 2} = \frac{2}{3}$$

**Practice Exericse No. 30**

Recuce these fractions to lowest terms.

| | | | |
|---|---|---|---|
| 1. $\frac{6}{16}$ | 6. $\frac{24}{40}$ | 11. $\frac{20}{50}$ | 16. $\frac{30}{54}$ |
| 2. $\frac{27}{72}$ | 7. $\frac{16}{100}$ | 12. $\frac{16}{36}$ | 17. $\frac{30}{64}$ |
| 3. $\frac{8}{32}$ | 8. $\frac{24}{36}$ | 13. $\frac{20}{25}$ | 18. $\frac{18}{54}$ |
| 4. $\frac{14}{28}$ | 9. $\frac{14}{16}$ | 14. $\frac{12}{56}$ | 19. $\frac{18}{48}$ |
| 5. $\frac{22}{33}$ | 10. $\frac{15}{45}$ | 15. $\frac{15}{50}$ | 20. $\frac{28}{56}$ |

## CHANGING WHOLE NUMBERS TO FRACTIONS

In using fractions in a problem requiring addition or subtraction, it is often necessary to combine them with whole numbers. In such instances, it will often be helpful to *change the whole numbers into fractions*.

EXAMPLE: Change the whole number five to an improper fraction with six as the denominator.

METHOD:

(a) Write five as a fraction with one as a denominator $\frac{5}{1}$ because $\frac{5}{1}$ is $5 \div 1$ which is the same as 5.

(b) Raise the fraction to higher terms.

$$\frac{5 \times 6}{1 \times 6} = \frac{30}{6}$$ (Multiplying numerator and denominator by same number does not change value.)

You will recall that a whole number written together with a fraction such as $2\frac{3}{4}$ is termed a *mixed number*.

**Rule: To change a mixed number to an improper fraction,** *change the whole number part to a fraction with the same denominator as the fraction; then add the two numerators and place the sum over the denominator.*

EXAMPLE: Change $2\frac{3}{4}$ to an improper fraction.

METHOD: $\frac{2 \times 4}{1 \times 4} = \frac{8}{4}$ So $\frac{2}{1} = \frac{8}{4}$ and now $\frac{8}{4} + \frac{3}{4} = \frac{11}{4}$. We do this because fractions must have the same denominator before we can add or subtract them.

EXAMPLE: Change $6\frac{3}{5}$ to an improper fraction. Think, 5 times 6 is 30 plus $3 = \frac{33}{5}$. Similarly with $5\frac{2}{7}$: 7 times 5 = 35, plus 2 equals 37 over 7. $5\frac{2}{7} = \frac{37}{7}$

**Rule: To change an improper fraction to a whole or mixed number,** *divide the numerator by the denominator and place the remainder over the denominator.*

EXAMPLE: Change $\frac{41}{3}$ to a mixed number.

METHOD:

$$\frac{41}{3} \text{ means } 41 \div 3 \text{ or } 3\overline{)41}^{\,13\,2R}$$

The remainder 2 is part of the divisor 3, so we write it as $\frac{2}{3}$ and the answer is $13\frac{2}{3}$.

Try another:

EXAMPLE: Change $\frac{87}{4}$ to a mixed number.

METHOD:

$$4\overline{)87}^{\,21\,?R} \text{ remainder} = ? \text{ divisor is }?$$

The answer should be written $21\frac{?}{?}$

### Practice Exercise No. 31

Change the mixed numbers below to improper fractions.

| | | |
|---|---|---|
| 1. $5\frac{1}{3}$ | 6. $2\frac{3}{7}$ | 11. $16\frac{1}{6}$ |
| 2. $12\frac{3}{5}$ | 7. $4\frac{2}{3}$ | 12. $12\frac{2}{7}$ |
| 3. $3\frac{1}{3}$ | 8. $1\frac{4}{9}$ | 13. $13\frac{3}{7}$ |
| 4. $4\frac{3}{8}$ | 9. $5\frac{3}{4}$ | 14. $14\frac{1}{5}$ |
| 5. $12\frac{1}{6}$ | 10. $8\frac{2}{3}$ | 15. $22\frac{2}{5}$ |

Change the improper fractions below to whole or mixed numbers and reduce to lowest terms.

| | | |
|---|---|---|
| 16. $\frac{25}{3}$ | 21. $\frac{42}{6}$ | 26. $\frac{21}{4}$ |
| 17. $\frac{63}{5}$ | 22. $\frac{35}{10}$ | 27. $\frac{60}{8}$ |
| 18. $\frac{25}{8}$ | 23. $\frac{35}{7}$ | 28. $\frac{41}{10}$ |
| 19. $\frac{34}{2}$ | 24. $\frac{8}{5}$ | 29. $\frac{46}{11}$ |
| 20. $\frac{25}{4}$ | 25. $\frac{8}{3}$ | 30. $\frac{16}{3}$ |

## IMPORTANCE OF THE LEAST COMMON DENOMINATOR

We cannot add or subtract different kinds of items and get an answer equal to the sum or difference of the numerical values. For example, if we added 2 apples and 3 oranges and arrived at a sum of 5, people would ask: 5 of what?

Similarly, if we want to *add or subtract fractions or compare them, they must have the same denominators. We call this denominator a* **common denominator.** For this reason, it is most important in dealing with fractions to know how to find a denominator that will be the same for any group of fractions. This denominator should be the *smallest* one that fits the need. The name given to this *new* denominator is the **least common denominator,** written **LCD.**

## FINDING THE COMMON DENOMINATOR

(1) Reduce the fraction to lowest terms if not already in this form.

(2) Can one denominator, the largest, serve for both fractions? If so, raise the lower fraction.

EXAMPLE: Find a common denominator between $\frac{5}{6}$ and $\frac{2}{3}$ and express both fractions with this denominator.

METHOD: Can we raise $\frac{2}{3}$ to 6th? *Yes.* Do it.

$$\frac{2 \times 2 = 4}{3 \times 2 = 6}$$

ANS. $\frac{4}{6}$ and $\frac{5}{6}$

EXAMPLE: Find the common denominator between $\frac{10}{16}$ and $\frac{3}{4}$.

METHOD: First reduce $\frac{10}{16}$ by dividing the numerator and denominator by 2.

$$\frac{10 \div 2 = 5}{16 \div 2 = 8}$$

Now raise $\frac{3}{4}$ to 8ths.

$$\frac{3 \times 2 = 6}{4 \times 2 = 8}$$

ANS. $\frac{5}{8}$ and $\frac{6}{8}$

If neither of the denominators can serve as a common denominator then multiply each denominator by a *different* number so they are the same.

EXAMPLE: Find the common denominator for $\frac{1}{2}$ and $\frac{1}{3}$ and express both fractions with this denominator.

METHOD: We cannot raise $\frac{1}{2}$ to thirds, but if we multiply 2 (the smallest) by 3 and 3 (the largest) by 2, both results will be 6. So raise both fractions to 6ths.

$$\frac{1 \times 3 = 3}{2 \times 3 = 6} \qquad \frac{1 \times 2 = 2}{3 \times 2 = 6}$$

ANS. $\frac{3}{6}$ and $\frac{2}{6}$

NOTE: *When choosing the numbers by which to multiply we must be sure to use the smallest numbers that will make both denominators the same so when we are finished we will have the* **least common denominator,** *or, as its called, the* **LCD.** *This is especially true when we have to find a common denominator for more than one fraction.*

## FINDING THE LEAST COMMON DENOMINATOR

EXAMPLE: What is the LCD of $\frac{1}{2}$, $\frac{1}{3}$, and $\frac{1}{4}$?

METHOD: Examine the denominators. Could any one of them serve as the denominator for all three fractions? No.

Try doubling the largest 4 $\times$ 2 or 8. Is 3 evenly divisible into 8? No. Try 4 $\times$ 3 or 12. Are 2 and 3 evenly divisible into 12? Yes. Then $\boxed{12}$ is the LCD.

### *Steps to Find a Common Denominator*

From the examples we can formulate these steps:

(1) Examine the fractions to see if one of them can serve as the common denominator as in $\frac{1}{3}$, $\frac{1}{2}$, and $\frac{1}{6}$.

(2) If none of the fractions has a common denominator, try doubling the greatest denominator. If that does not fit, try tripling the greatest denominator and so on until you find the right number.

As an alternative or substitute method for the second step, we may multiply the denominators by each other to arrive at a common denominator. The resultant common denominator may have to be reduced to the LCD.

### Practice Exercise No. 32

Find the least common denominators in the examples below.

**1.** $\frac{1}{3}$ and $\frac{1}{12}$    **3.** $\frac{2}{3}$ and $\frac{1}{5}$    **5.** $\frac{1}{4}$ and $\frac{5}{6}$

**2.** $\frac{2}{5}$ and $\frac{1}{2}$    **4.** $\frac{2}{9}$ and $\frac{1}{2}$    **6.** $\frac{2}{3}$ and $\frac{5}{8}$

7. $\frac{5}{6}, \frac{3}{8}, \frac{1}{2}$     9. $\frac{5}{6}, \frac{5}{12}, \frac{1}{8}$     11. $\frac{1}{9}, \frac{1}{7}, \frac{2}{3}$

8. $\frac{1}{2}, \frac{5}{8}, \frac{3}{4}$     10. $\frac{2}{3}, \frac{1}{10}, \frac{3}{5}$     12. $\frac{1}{5}, \frac{3}{9}, \frac{2}{7}$

## ADDITION OF FRACTIONS

Fractions are called *like fractions* when they have the same denominator.

**Rule: To add like fractions,** *add the numerators and place the sum over the denominator.*

EXAMPLE: Add $\frac{1}{8} + \frac{3}{8} + \frac{5}{8}$

$$\frac{1 + 3 + 5}{8} = \frac{9}{8} \text{ or } 1\frac{1}{8} \text{ Ans.}$$

EXAMPLE: Add $\frac{1}{5} + \frac{2}{5} + \frac{2}{5}$

$$\frac{1 + 2 + 2}{5} = \frac{5}{5} \text{ or } 1 \text{ Ans.}$$

**Rule: To add unlike fractions,** *change the given fractions to their equivalent fractions, all having the same denominator. Then add the numerators and place the sum over the common denominator.*

EXAMPLE: Add $\frac{5}{8} + \frac{5}{6}$

What is the LCD? Think 8, 16, $\boxed{24}$.

$\frac{5}{8} = \frac{?}{24} = $ Think $24 \div 8 = 3$;

$$5 \times 3 = 15 = \frac{15}{24}$$

$\frac{5}{6} = \frac{?}{24} = $ Think $24 \div 6 = 4$;

$$5 \times 4 = 20 = \frac{20}{24}$$

$$\frac{15 + 20}{24} = \frac{35}{24} = 1\frac{11}{24} \text{ Ans.}$$

EXAMPLE: Add $\frac{2}{3} + \frac{3}{5} + \frac{1}{2}$

What is the LCD? Multiply $5 \times 3 \times 2 = 30$. (See alternative method of finding a common denominator.)

$\frac{2}{3} = \frac{?}{30}$ Think $30 \div 3 = 10$;

$$2 \times 10 = 20 \text{ or } \frac{20}{30}$$

$\frac{3}{5} = \frac{?}{30}$ Think $30 \div 5 = 6$;

$$3 \times 6 = 18 \text{ or } \frac{18}{30}$$

$\frac{1}{2} = \frac{?}{30}$ Think $30 \div 2 = 15$;

$$1 \times 15 = 15 \text{ or } \frac{15}{30}$$

$$\frac{20 + 18 + 15}{30} = \frac{53}{30} = 1\frac{23}{30} \text{ Ans.}$$

**Rule: To add mixed numbers,** *treat the fractions separately, then add the results to the whole numbers.*

EXAMPLE: Add: $5\frac{3}{4} + 7\frac{5}{8}$. What is the LCD?

$$\begin{array}{r} 5\frac{3}{4} = \quad\; 5\frac{6}{8} \\ + 7\frac{5}{8} = \; + 7\frac{5}{8} \\ \hline 12\frac{11}{8} = 12 + 1\frac{3}{8} = 13\frac{3}{8} \text{ Ans.} \end{array}$$

EXAMPLE: Add $12\frac{2}{3} + \frac{16}{5} + 18\frac{5}{6}$

What is the LCD. Think 6, 12, 18, 24, $\boxed{30}$. That's it!

$$12\frac{2}{3} = 12\frac{20}{30}$$

$$\frac{16}{5} = \; 3\frac{1}{5} = \; 3\frac{6}{30}$$

$$18\frac{5}{6} = 18\frac{25}{30}$$
$$\overline{33\frac{51}{30}} = 33 + 1\frac{21}{30} = 34\frac{21}{30}$$

$$\frac{21}{30} \text{ reduces to } \frac{?}{10}$$

**Practice Exercise No. 33**

Find the sums of the following fractions and reduce your answers to lowest terms.

1. $\frac{3}{7} + \frac{4}{7} + \frac{6}{7}$      9. $8\frac{5}{12} + 9\frac{2}{8}$

2. $5\frac{3}{4} + \frac{3}{4}$     10. $4\frac{4}{5} + 7\frac{6}{10} + 6\frac{4}{15}$

3. $\frac{7}{12} + 4\frac{11}{12}$     11. $8\frac{3}{4} + \frac{2}{3} + 3\frac{3}{12}$

4. $5\frac{5}{6} + 8\frac{5}{6}$     12. $6\frac{1}{3} + 9\frac{3}{8} + 8\frac{5}{12}$

5. $4\frac{1}{3} + 12\frac{1}{3} + 7\frac{1}{3}$     13. $7\frac{1}{4} + 8\frac{3}{5}$

6. $\frac{1}{3} + \frac{5}{6}$     14. $22\frac{5}{8} + 12\frac{5}{12}$

7. $2\frac{1}{12} + 5\frac{1}{4}$     15. $4\frac{3}{4} + 5\frac{1}{6} + 6\frac{2}{3}$

8. $3\frac{3}{5} + 5\frac{1}{10}$

## SUBTRACTION OF FRACTIONS

In order to subtract one fraction from another we must, of course, have *like fractions*.

**Rule: To subtract like fractions,** *subtract the numerators and place the difference over the denominator.*

EXAMPLE: Subtract $\frac{2}{5}$ from $\frac{4}{5}$

$$\frac{4-2}{5} = \frac{2}{5}$$

$$\frac{5}{6} - \frac{3}{6} = ? \quad \frac{7}{8} - \frac{1}{8} = ? \quad \frac{8}{11} - \frac{3}{11} = ?$$

**Rule: To subtract unlike fractions,** *find their LCD, and then find the difference between the new numerators.*

EXAMPLE: Subtract $\frac{3}{4}$ from $\frac{7}{8}$.

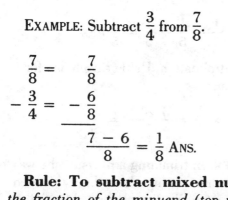

$$\frac{7}{8} = \frac{7}{8}$$
$$-\frac{3}{4} = -\frac{6}{8}$$
$$\frac{7-6}{8} = \frac{1}{8} \text{ Ans.}$$

**Rule: To subtract mixed numbers,** *the fraction of the minuend (top number), must be larger than the fraction of the subtrahend (the bottom number), then proceed to find the difference.*

EXAMPLE 1: Subtract $2\frac{1}{2}$ from 4.

$$\begin{array}{rcl} 4 &=& 3\frac{2}{2} \\ -\ 2\frac{1}{2} &=& -\ 2\frac{1}{2} \\ \hline && 1\frac{1}{2} \text{ Ans.} \end{array}$$

EXPLANATION: We change 4 to $3\frac{2}{2}$ by exchanging the 1 in the minuend for $\frac{2}{2}$ to make the top fraction larger and at the same time find a *common* denominator.

EXAMPLE 2: Subtract $2\frac{1}{2}$ from $5\frac{2}{3}$.

$$\begin{array}{rcl} 5\frac{2}{3} &=& 5\frac{4}{6} \\ -\ 2\frac{1}{2} &=& -\ 2\frac{3}{6} \\ \hline && 3\frac{1}{6} \text{ Ans.} \end{array}$$

EXPLANATION: The only change needed was to find the common denominator 6, and raise both fractions to 6ths.

EXAMPLE 3: Subtract $2\frac{3}{4}$ from $5\frac{1}{4}$.

$$\begin{array}{rcl} 5\frac{1}{4} &=& 4\frac{5}{4} \\ -\ 2\frac{3}{4} &=& -\ 2\frac{3}{4} \\ \hline 2\frac{2}{4} &=& 2\frac{1}{2} \text{ Ans.} \end{array}$$

EXPLANATION: Since $\frac{3}{4}$ is greater than $\frac{1}{4}$, it was necessary to change $5\frac{1}{4}$ to $4\frac{5}{4}$ by exchanging a one from the 5 to make $\frac{4}{4}$ and adding it to $\frac{1}{4}$ to give $4\frac{5}{4}$.

EXAMPLE 4: Subtract $2\frac{3}{4}$ from $5\frac{1}{6}$.

$$\begin{array}{rcccl} 5\frac{1}{6} &=& 5\frac{2}{12} &=& 4\frac{14}{12} \\ -\ 2\frac{3}{4} &=& -\ 2\frac{9}{12} &=& -\ 2\frac{9}{12} \\ \hline &&&& 2\frac{5}{12} \text{ Ans.} \end{array}$$

EXPLANATION: First we raised $\frac{3}{4}$ and $\frac{1}{6}$ to like fractions. Next, it was necessary to exchange 1 from the 5 to make $\frac{12}{12}$, and added it to the $\frac{2}{12}$ to equal $4\frac{14}{12}$ making possible the subtraction of fractions.

### Practice Exercise No. 34

Do the subtraction of fractions below and reduce your answers to lowest terms.

1. $9\frac{3}{4} - 9\frac{1}{4} =$

2. $5\frac{1}{4} - 5 =$

3. $7\frac{5}{6} - \frac{1}{3} =$

4. $5\frac{3}{8} - 3\frac{1}{3} =$

5. $10\frac{1}{2} - 2\frac{2}{5} =$

6. $10 - \frac{5}{6} =$

7. $15 - 7\frac{1}{4} =$

8. $12\frac{1}{6} - 11\frac{5}{6} =$

9. $15\frac{2}{7} - 5\frac{3}{7} =$

10. $15\frac{15}{16} - 4\frac{7}{8} =$

11. $10\frac{1}{4} - 3\frac{11}{12} =$

12. $8\frac{2}{3} - 3\frac{7}{9} =$

13. $5\frac{1}{2} - 2\frac{2}{3} =$

14. $8\frac{5}{6} - 3\frac{7}{8} =$

15. $10\frac{1}{4} - 6\frac{2}{3} =$

16. $\frac{7}{8} - \frac{1}{2} - \frac{1}{4} =$

### Practice Exercise No. 35

The problems below will test your ability to add and subtract fractions. Work carefully remembering what you have learned about finding common denominators.

**1.** Lucille's mother is taking an overseas airplane trip. She is allowed 66 pounds for luggage. Her large bag weighs $27\frac{3}{8}$ lb., and her small bag $12\frac{3}{4}$ lb. How many pounds is she below her limit?

**2.** Hank the center on the basketball team is $67\frac{1}{2}$ inches tall. His rival for the position is $65\frac{1}{4}$ inches tall. How much taller is Hank?

**3.** Ricky is expected to practice his accordion lessons 5 hours per week. Monday he practiced $1\frac{1}{4}$ hours, Tuesday $\frac{3}{4}$ hour, Wednesday $\frac{1}{2}$ hour. On Thursday he did not practice. On Friday he put in $1\frac{1}{2}$ hours. How much more time must he practice to make up the required total?

**4.** John and his father decided to wallpaper their hobby room. They estimated the 4 walls would require the following amounts of wallpaper: $2\frac{1}{2}$ rolls, $3\frac{5}{8}$ rolls, $1\frac{3}{4}$ rolls and $2\frac{1}{8}$ rolls. How many rolls would they need in all?

**5.** Ruth decided to bake. She had 6 cups of flour. Her recipes called for $2\frac{1}{2}$ cups for cookies, $1\frac{1}{2}$ cups for a pie and $3\frac{1}{4}$ cups for a cake. How many more cups of flour will she need?

**6.** Helen sells eggs during the summer. She sold $12\frac{1}{2}$ dozen the first week, $9\frac{1}{4}$ dozen the second week and $11\frac{3}{4}$ the third week. How many dozen eggs must she sell the fourth week in order to have sold 50 dozen at the end of four weeks?

**7.** George bought a $3\frac{1}{4}$ horsepower (H.P.) outboard motor for his boat. Harry had a 7 H.P. motor while Jim was using a $10\frac{1}{2}$ H.P. motor. (a) How much greater H.P. was Harry's than George's? (b) How much greater was Jim's than Harry's?

**8.** Alex spent $\frac{1}{2}$ hour on his spelling studies, $\frac{1}{3}$ hour on social studies and $\frac{3}{4}$ hour on arithmetic. How much time did he spend on his homework?

**9.** Henry and Joe went fishing. Henry caught a bass that weighed $6\frac{3}{8}$ lbs. Joe caught one that weighed $8\frac{1}{4}$ lb. How much heavier was Joe's fish?

**10.** Jack is expected to do 10 hours of chores around the house and grounds per week, beginning Monday and ending Friday, during the summer to earn his $15.00 allowance. During the first week he worked $1\frac{1}{2}$ hours on Monday, $2\frac{1}{4}$ hours Tuesday, $1\frac{2}{3}$ hours Wednesday, $\frac{3}{4}$ hour Thursday and $1\frac{1}{3}$ hours Friday. His father decided to deduct from Jack's allowance an amount proportionate to the time he failed to put in during the week. How much allowance did Jack receive this first week?

## MULTIPLICATION OF FRACTIONS

### Multiplying a Fraction by a Whole Number

PROBLEM: Lewis had a package of chewing gum containing five sticks of gum. He was permitted to take one stick of gum a day. What part of the package of gum did Lewis use at the end of three days?

METHOD: 1 stick of gum represents $\frac{1}{5}$ of the package.

By addition:

$$\frac{1}{5} + \frac{1}{5} + \frac{1}{5} = \frac{3}{5} \text{ ANS.}$$

By multiplication: Using $\frac{1}{5}$ each day for 3 days, means

$$3 \times \frac{1}{5} = \frac{3 \times 1}{5} = \frac{3}{5} \text{ ANS.}$$

PROBLEM: In trimming a playsuit she was making in her sewing class, Laury needed 12 pieces of ribbon, each $\frac{2}{3}$ yd. long. How long a piece of ribbon did she have to buy?

METHOD: If each piece is $\frac{2}{3}$ yd. and 12 pieces are needed, then she requires

$$12 \times \frac{2}{3} = \frac{12 \times 2}{3} = \frac{24}{3} = 8 \text{ yds. Ans.}$$

NOTE: In multiplying fractions by whole numbers, you may write the whole number in fraction form (as an improper fraction) as in the following examples:

EXAMPLE: Multiply $5 \times \frac{1}{6}$

$$\frac{5}{1} \times \frac{1}{6} = \frac{5}{6} \text{ Ans.}$$

EXAMPLE: Multiply $2 \times \frac{2}{3}$

$$\frac{2}{1} \times \frac{2}{3} = \frac{4}{3} = 1\frac{1}{3} \text{ Ans.}$$

**Rule: To multiply a fraction by a whole number,** *multiply the* numerator *of the fraction by the whole number and place the product over the denominator.*

### Multiplying a Whole Number by a Fraction

PROBLEM: In a litter of 16 rabbits, it was expected that $\frac{3}{4}$ would be white. How many white rabbits did they expect to find in the litter?

FIGURE 6.

METHOD: Saying $\frac{3}{4}$ of the litter would be white, is the same as saying, out of every four rabbits, three would be white.

We can show this in an illustration (Figure 6). Here we see groups of four. Since three out of every four are white, the answer would be $3 \times 4$ or 12.

By this you can see that:

$$\frac{3}{4} \text{ of 16 is the same as}$$

$$\frac{3}{4} \times \frac{16}{1} = \frac{48}{4} = 12 \text{ Ans.}$$

OBSERVE: To understand the multiplication of fractions it is important to recognize that:

(a) Multiplying by a fraction means taking a part of it.

(b) When you see a fraction followed by the word **"of"** it means the same as times.

(c) Since a proper fraction is *less than one*, any number multiplied by a proper fraction, will have a product lower than the original number.

(d) The denominator of any whole number is one.

When you divide a number by 1, does it change the value?

EXAMPLE 1: Multiply $\frac{2}{3} \times 1$

$$\frac{2}{3} \times \frac{1}{1} = \frac{2}{3} \text{ Ans.}$$

EXAMPLE 2: Multiply $\frac{2}{3} \times 9$

$$\frac{2}{3} \times \frac{9}{1} = \frac{18}{3} = 6 \text{ Ans.}$$

EXAMPLE 3: $\frac{5}{6}$ of $12 =$

$$\frac{5}{6} \times \frac{12}{1} = \frac{60}{6} = 10 \text{ Ans.}$$

EXAMPLE 4: $\frac{2}{3}$ of 21 =

$$\frac{2}{3} \times \frac{21}{1} = \frac{42}{3} = 14 \text{ Ans.}$$

Is there a difference between these?

$$3 \times \frac{2}{3} \quad \text{and} \quad \frac{2}{3} \times 3$$

The answer is no. Therefore complete the wording of this rule by filling in the spaces.

**Rule: To multiply a whole number by a fraction,** *multiply the _____ of the fraction by the _____ _____ and place the _____ over the _____ .*

### Practice Exercise No. 36

Multiply the fractions below. Reduce your answers to lowest terms.

1. $2 \times \frac{1}{2}$     7. $8 \times \frac{2}{5}$     13. $\frac{1}{16} \times 30$

2. $\frac{1}{3} \times 2$     8. $\frac{3}{4}$ of 12     14. $\frac{9}{10}$ of 5

3. $3 \times \frac{1}{3}$     9. $4 \times \frac{5}{7}$     15. $8 \times \frac{2}{9}$

4. $\frac{4}{5} \times 16$     10. $\frac{5}{8}$ of 3     16. $\frac{2}{3} \times 7$

5. $4 \times \frac{2}{3}$     11. $2 \times \frac{9}{16}$

6. $5 \times \frac{2}{7}$     12. $\frac{1}{6} \times 8$

### Multiplying Whole Numbers by Mixed Numbers

PROBLEM: A box of stainless-steel screws weighs $2\frac{3}{4}$ ounces. How much will 5 boxes weigh?

METHOD: Multiply $5 \times 2\frac{3}{4}$.

Change $2\frac{3}{4}$ to an improper fraction, $\frac{11}{4}$

$$\frac{5}{1} \times \frac{11}{4} = \frac{55}{4} = 13\frac{3}{4} \text{ ounces Ans.}$$

A variation of this method is useful when the whole numbers are large. For example, in the previous problem:

Change the mixed number into a whole number and a fraction. Then multiply separately.

$$5 \times 2\frac{3}{4} = 5 \times 2 = 10$$

$$\text{and } \frac{5}{1} \times \frac{3}{4} = \frac{15}{4} = 3\frac{3}{4}$$

$$10 + 3\frac{3}{4} = 13\frac{3}{4} \text{ Ans.}$$

When working with larger whole numbers the problem would be set up as follows:

EXAMPLE: Multiply $24 \times 18\frac{1}{5}$

$$24 \times 18 = 432 \text{ and } 24 \times \frac{1}{5} = \frac{24}{5} = 4\frac{4}{5}$$

$$432 + 4\frac{4}{5} = 436\frac{4}{5} \text{ Ans.}$$

### Practice Exercise No. 37

Using either of the methods described above, multiply the fractions below.

1. $1\frac{1}{2} \times 3$     5. $6 \times 6\frac{1}{6}$     9. $7 \times 24\frac{1}{2}$

2. $4 \times \frac{2}{3}$     6. $6 \times 2\frac{3}{8}$     10. $28 \times 1\frac{1}{7}$

3. $8 \times 6\frac{1}{4}$     7. $7\frac{1}{4} \times 8$     11. $42 \times 16\frac{2}{3}$

4. $10\frac{1}{2} \times 2$     8. $24 \times 10\frac{2}{3}$     12. $64 \times 32\frac{3}{8}$

### Multiplying a Fraction by a Fraction

PROBLEM: Dick's mother had $\frac{1}{2}$ of a pie left. He came home with 2 friends. The 3 boys shared the rest of the pie equally. What part of the whole pie did Dick get?

METHOD: Look at the illustrations below.

$\frac{1}{3}$ of this

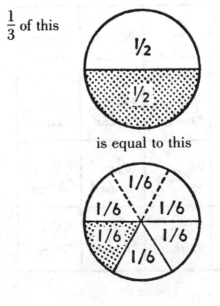

is equal to this

The whole pie was originally divided into 6 parts. Since $\frac{1}{3}$ of $\frac{1}{2}$ pie is the same as $\frac{1}{6}$ of the whole pie, then

$$\frac{1}{3} \text{ of } \frac{1}{2} = \frac{1}{3} \times \frac{1}{2} = \frac{1}{6}$$

REMEMBER: In the multiplication of fractions, **of** means times.

PROBLEM: In a previous example we referred to a litter of 16 rabbits of which $\frac{3}{4}$ were white; suppose we said $\frac{1}{3}$ of the white rabbits had extra large ears. How many rabbits with extra large ears would there be?

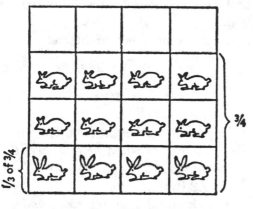

FIGURE 7.

METHOD: We are asking, how much is $\frac{1}{3}$ of $\frac{3}{4}$. From the illustration (Figure 7), we see that $\frac{3}{4}$ of the 16 rabbits is 12 and $\frac{1}{3}$ of those 12 is 4. We see then that 4 long-eared rabbits of the total group of 16 is $\frac{4}{16}$ or $\frac{1}{4}$.

By multiplication:

$$\frac{1}{3} \text{ of } \frac{3}{4} \text{ is } \frac{1}{3} \times \frac{3}{4} = \frac{3}{12} \text{ or } \frac{1}{4} \text{ Ans.}$$

**Rule: To multiply a fraction by a fraction,** *multiply the numerators to get the new numerator. Multiply the denominators to get the new denominator. Change the fraction to simplest form.*

EXAMPLE 1: Multiply $\frac{2}{3} \times \frac{2}{5}$

$$\frac{2}{3} \times \frac{2}{5} = \frac{2 \times 2}{3 \times 5} = \frac{4}{15} \text{ Ans.}$$

EXAMPLE 2: $\frac{3}{4}$ of $\frac{3}{5}$

$$\frac{3}{4} \times \frac{3}{5} = \frac{3 \times 3}{4 \times 5} = \frac{9}{20} \text{ Ans.}$$

### Multiplying Mixed Numbers

PROBLEM: Alice wants to increase a recipe for four people to one which will serve six people. This means multiplying by $1\frac{1}{2}$. How much sugar will she need if the original recipe calls for $2\frac{2}{3}$ cups?

METHOD: This means taking $2\frac{2}{3}$ cups $1\frac{1}{2}$ times or $2\frac{2}{3} \times 1\frac{1}{2}$. Change both mixed numbers to improper fractions. $2\frac{2}{3} = \frac{8}{3}$ and $1\frac{1}{2} = \frac{3}{2}$.

Multiply
$$\frac{8}{3} \times \frac{3}{2} = \frac{8 \times 3}{3 \times 2} = \frac{24}{6} = 4 \text{ Ans.}$$

EXAMPLE: $3\frac{1}{5} \times 2\frac{1}{3} =$

Converting to improper fractions

$$3\frac{1}{5} = \frac{16}{5} \text{ and } 2\frac{1}{3} = \frac{7}{3};$$

$$\frac{16}{5} \times \frac{7}{3} = \frac{16 \times 7}{5 \times 3} = \frac{112}{15} = 7\frac{7}{15} \text{ Ans.}$$

## A SHORT WAY OF MULTIPLYING BY FRACTIONS

To shorten the work in multiplying fractions, we can apply some principles we learned before.

We know that *we can divide the numerator and denominator of a fraction by the same number without changing its value.*

See how it applies in this problem.

EXAMPLE 1: $\dfrac{3}{4} \times \dfrac{4}{5} = \dfrac{3 \times \overset{1}{\cancel{4}}}{\underset{1}{\cancel{4}} \times 5} = \dfrac{3}{5}$ Ans.

Before multiplying, we divided the numerator and denominator by four. This simplified the multiplication.

This process has recently been termed in some school systems as "division before multiplication." In most books and many school systems it is called cancellation.

The actual process is *division*.

The *effect* is that of *cancelling* or *reducing* numbers in the numerator and denominator before multiplying.

Whatever you call it, the process is a helpful time saver. It may be used in many places and can be used more than once in the same problem. Observe how it shortens the work in this problem.

EXAMPLE 2: $\dfrac{16}{24} \times \dfrac{24}{32}$

The short way:

$$\frac{\cancel{16}}{\cancel{24}} \times \frac{\cancel{24}}{\cancel{32}} = \frac{1}{2} \text{ ANS.}$$

The long way:

$\dfrac{16 \times 24}{24 \times 32} = \dfrac{384}{768}$  This fraction can be reduced to _____ ?

Divide the numerator and denominator by 24 to get $\frac{16}{32}$. Then divide both numerator and denominator by 16 to get $\frac{1}{2}$. You see this is effectively what we did when we *cancelled* the 16 and 24 out before we multiplied.

EXAMPLE 3: Multiply $\dfrac{3}{8} \times \dfrac{5}{6} \times \dfrac{2}{15}$

Short way:

$$\frac{\cancel{3} \times \cancel{5} \times \cancel{2}}{8 \times \cancel{6} \times \cancel{15}} = \frac{1}{24} \text{ ANS.}$$

Long way:

$\dfrac{3 \times 5 \times 2}{8 \times 6 \times 15} = \dfrac{30}{720}$ reduces to what?

Do the examples below. Use "division before multiplication" where possible.

1. $\frac{3}{8} \times \frac{1}{2}$          7. $\frac{5}{6} \times \frac{9}{10}$

2. $\frac{2}{7} \times \frac{1}{8}$          8. $\frac{2}{5} \times 1\frac{1}{6}$

3. $\frac{5}{9} \times \frac{1}{2}$          9. $\frac{2}{3} \times 7\frac{1}{2}$

4. $\frac{3}{8} \times \frac{2}{3}$          10. $\frac{3}{4} \times 2\frac{1}{3}$

5. $\frac{3}{4} \times \frac{11}{12}$        11. $1\frac{1}{5} \times 7\frac{1}{2}$

6. $\frac{3}{16} \times \frac{2}{9}$         12. $1\frac{3}{8} \times 3 \times 1\frac{1}{3}$

## DIVISION OF FRACTIONS

The method for dividing by fractions may best be explained by the following situation:

Compare these two problems.

$$2\overline{)10} = 5$$
$$\frac{1}{2} \times 10 = 5$$

The same answer is obtained when we *divide 10 by 2* as when we *multiply 10 by $\frac{1}{2}$*.

The number $\frac{1}{2}$ is the "reciprocal" of the number 2 (which is $\frac{2}{1}$).

**A reciprocal** *is an inverted number.* We say $\frac{1}{5}$ is the reciprocal of 5; $\frac{1}{3}$ is the reciprocal of 3; $\frac{5}{2}$ is the reciprocal of $\frac{2}{5}$; 5 is the reciprocal of $\frac{1}{5}$.

**Rule: To divide by a fraction,** *multiply by the reciprocal.*

To say this another way:

**Rule: To divide in a problem containing a fraction,** *invert the* divisor *and then multiply.*

In all division examples, it is important that you *learn to identify the divisor,* because the *divisor* is the number to be inverted.

$$10 \div 2 = 5 \text{ but } 2 \div 10 = \frac{2}{10} \text{ or } \frac{1}{5}.$$

Again

$$\frac{2}{3} \div \frac{3}{4} \text{ is not the same as } \frac{3}{4} \div \frac{2}{3}$$

### Dividing a Fraction by a Whole Number

PROBLEM: The apple pie in Al's restaurant had been cut into eight pieces. There were five pieces left. That would be $\frac{5}{8}$ of the pie. If you and four friends each ordered one piece, what part of the original pie would each of you be getting?

METHOD: We have $\frac{5}{8}$ to divide by 5. The *divisor* here is a whole number.

$$\frac{5}{8} \div 5 \text{ or } \frac{5}{8} \div \frac{5}{1} \text{ is the same as } \frac{5}{8} \times \frac{1}{5}$$

$$\frac{\overset{1}{5}}{8} \times \frac{1}{\underset{1}{5}} = \frac{1}{8} \text{ ANS.} \qquad \text{(Invert the divisor and multiply.)}$$

Now take $\frac{5}{6}$ of a pie and divide it among 5 people. How much would each one get? Look at the illustration.

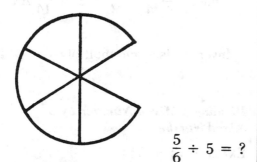

$$\frac{5}{6} \div 5 = ?$$

EXAMPLE: $\frac{4}{5} \div 4 = ?$

$$\frac{4}{5} \div \frac{4}{1} = \frac{\overset{1}{4}}{5} \times \frac{1}{\underset{1}{4}} = \frac{1}{5} \text{ ANS.}$$

### Dividing a Whole Number by a Fraction

PROBLEM: Bart was repairing his boat. He bought 12 runnings yards of fiberglass material which was to be cut into pieces $\frac{3}{8}$ yd. long for patching the lap seams. How many such pieces could he get out of it?

METHOD: We have to divide 12 by $\frac{3}{8}$. The divisor here is a fraction

$$12 \div \frac{3}{8} \text{ or } \frac{12}{1} \div \frac{3}{8} \text{ is the same as } \frac{12}{1} \times \frac{8}{3}$$

$$\frac{\overset{4}{12}}{1} \times \frac{8}{\underset{1}{3}} = 32 \text{ ANS.} \qquad \text{(Invert the divisor and multiply.)}$$

Look at the section of a ruler pictured here. How many $\frac{1}{4}$ inch divisions are there in the 2 inch piece?

$$2 \div \frac{1}{4} = ?$$

EXAMPLE: $9 \div \frac{3}{5} = ?$

$$\frac{9}{1} \div \frac{3}{5} = \frac{\overset{3}{9}}{1} \times \frac{5}{\underset{1}{3}} = 15 \text{ ANS.}$$

### Practice Exercise No. 39

Find the quotients in the problems which follow.

1. $\frac{5}{6} \div 4$     5. $\frac{2}{3} \div 14$    9. $14 \div \frac{2}{7}$
2. $\frac{2}{3} \div 6$    6. $4 \div \frac{1}{2}$    10. $24 \div \frac{3}{8}$
3. $\frac{5}{6} \div 5$    7. $5 \div \frac{1}{10}$    11. $15 \div \frac{3}{4}$
4. $\frac{7}{8} \div 4$    8. $15 \div \frac{5}{8}$    12. $48 \div \frac{2}{3}$

### Dividing a Fraction by a Fraction

PROBLEM: Barbara had $\frac{3}{4}$ yards of silk ribbon for bows to put on her blouse. Each bow needed $\frac{1}{12}$ yd. of ribbon. How many bows could she make from the $\frac{3}{4}$ yd.?

METHOD: We have to divide $\frac{3}{4}$ by $\frac{1}{12}$. The divisor is the fraction $\frac{1}{12}$

$$\frac{3}{4} \div \frac{1}{12} \text{ is the same as } \frac{3}{\cancel{4}} \times \frac{\cancel{12}^{3}}{1} = 9 \text{ ANS.}$$

In the section of the ruler pictured above, how many $\frac{1}{8}$ inch divisions are there in a $\frac{1}{2}$ inch section? Count them.

Now divide $\frac{1}{2} \div \frac{1}{8} = ?$

EXAMPLE: $\dfrac{2}{9} \div \dfrac{1}{3} = \dfrac{2}{\cancel{9}_3} \times \dfrac{\cancel{3}^1}{1} = \dfrac{2}{3}$ ANS.

### Dividing When There Are Mixed Numbers

The same methods that are used for division of proper fractions apply in division of mixed numbers. However, we must first change the mixed numbers to improper fractions. Note the procedure in these examples.

### Dividing a Whole Number by a Mixed Number

EXAMPLE: $15 \div 1\frac{2}{3}$; change $1\frac{2}{3}$ to $\frac{5}{3}$,

then $\dfrac{15}{1} \div \dfrac{5}{3} = \dfrac{\cancel{15}^3}{1} \times \dfrac{3}{\cancel{5}_1} = 9$ ANS.

Invert the divisor and multiply as usual.

### Dividing a Mixed Number by a Whole Number

EXAMPLE: $5\frac{1}{3} \div 8$; change $5\frac{1}{3}$ to $\frac{16}{3}$,

then $\dfrac{16}{3} \div \dfrac{8}{1} = \dfrac{\cancel{16}^2}{3} \times \dfrac{1}{\cancel{8}_1} = \dfrac{2}{3}$

Invert the divisor and multiply.

### Dividing a Mixed Number by a Proper Fraction

EXAMPLE: $6\frac{2}{3} \div \frac{2}{3}$; change $6\frac{2}{3}$ to $\frac{20}{3}$,

then $\dfrac{20}{3} \div \dfrac{2}{3} = \dfrac{\cancel{20}^{10}}{\cancel{3}_1} \times \dfrac{\cancel{3}^1}{\cancel{2}_1} = 10$ ANS.

Invert divisor and multiply.

### Dividing a Proper Fraction by a Mixed Number

EXAMPLE: $\frac{3}{8} \div 1\frac{3}{4}$; change $1\frac{3}{4}$ to $\frac{7}{4}$,

then $\dfrac{3}{8} \div \dfrac{7}{4} = \dfrac{3}{\cancel{8}_2} \times \dfrac{\cancel{4}^1}{7} = \dfrac{3}{14}$ ANS.

Invert divisor and multiply.

### Dividing a Mixed Number by a Mixed Number

EXAMPLE: $5\frac{1}{3} \div 1\frac{1}{3}$

Change $5\frac{1}{3}$ to $\frac{16}{3}$; change $1\frac{1}{3}$ to $\frac{4}{3}$

$\dfrac{16}{3} \div \dfrac{4}{3} = \dfrac{\cancel{16}^4}{\cancel{3}_1} \times \dfrac{\cancel{3}^1}{\cancel{4}_1} = 4$ ANS.

Invert divisor and multiply.

## CHECKING DIVISION OF FRACTIONS

Examples involving division of fractions are checked in the same manner as examples involving division of whole numbers.

Since: Dividend ÷ divisor = quotient
Check: Quotient × divisor = dividend

EXAMPLE:

$$15 \div \frac{5}{3} = 9$$

Dividend ÷ divisor = quotient

Check:

$$9 \times \frac{5}{3} = \frac{45}{3} = 15$$

Quotient × divisor = dividend

Try this check with two fractions

$$\frac{2}{9} \div \frac{1}{3} = \frac{2}{9} \times \frac{3}{1} = \frac{2}{3}$$

Dividend ÷ divisor = quotient

Check:

$$\frac{2}{3} \times \frac{1}{3} = \frac{2}{9}$$

Quotient × divisor = dividend

### Practice Exercise No. 40

Do the following examples involving division of varied fractions. Check your work carefully.

| | | |
|---|---|---|
| **1.** $2\frac{2}{5} \div \frac{3}{5}$ | **5.** $2\frac{7}{8} \div \frac{1}{2}$ | **9.** $7\frac{2}{7} \div \frac{3}{14}$ |
| **2.** $2\frac{3}{4} \div 5$ | **6.** $3\frac{3}{8} \div \frac{3}{4}$ | **10.** $6\frac{3}{12} \div \frac{3}{8}$ |
| **3.** $2\frac{7}{16} \div \frac{3}{8}$ | **7.** $5\frac{1}{4} \div 2\frac{3}{4}$ | **11.** $2\frac{4}{9} \div \frac{11}{12}$ |
| **4.** $2\frac{3}{4} \div 3\frac{1}{2}$ | **8.** $6\frac{1}{2} \div \frac{7}{8}$ | **12.** $6\frac{7}{8} \div 3\frac{3}{4}$ |

### Finding the Whole When a Fractional Part is Given

We shall now discuss the methods of finding the whole when a fractional part is given.

PROBLEM: If $\frac{1}{3}$ of the capacity of a gas tank is five gallons, what is the full capacity of the tank?

METHOD:

$$5 \div \frac{1}{3} = \frac{5}{1} \times \frac{3}{1} = 15 \text{ gal. ANS.}$$

Let's try a more difficult example of the same type.

PROBLEM: Harold spent \$4. This is $\frac{2}{3}$ of his weekly allowance. What is his weekly allowance?

METHOD: If \$4 is $\frac{2}{3}$, what is $\frac{1}{3}$?

$4 \div 2 = \$2$. This is $\frac{1}{3}$ of his allowance.

$$\$2 \div \frac{1}{3} = \frac{\$2}{1} \times \frac{3}{1} = \$6 \text{ ANS.}$$

or $4 \div \frac{2}{3}$

$$\frac{4}{1} \div \frac{2}{3} = \frac{4}{1} \times \frac{3}{2} = \frac{12}{2} = 6$$

Let's find a rule to fit what we did above.

**To find the whole, when a fractional amount is known,** *divide the amount by the fraction that is given.*

Can we do the problem above in one step?

$$\$4 \div \frac{2}{3} = \frac{\$4}{1} \times \frac{3}{2} = ?$$

EXAMPLE: If $\frac{2}{7}$ of a number is 6 what is the whole number?

$$6 \div \frac{2}{7} = \overset{3}{\cancel{6}} \times \frac{7}{\underset{1}{2}} = 21 \text{ ANS.}$$

To check: Take $\frac{2}{7}$ of 21. $\frac{2}{7} \times 21 = ?$

### Practice Exercise No. 41

The examples and problems which follow will test your ability to work with fractions. The first eight examples require you to determine the whole when a fractional part is given. What is the whole if:

1. $\frac{1}{3}$ of it is 15

2. $\frac{1}{12}$ of it is 7

3. $\frac{1}{2}$ of it is $6\frac{1}{2}$

4. $\frac{2}{3}$ of it is 12

5. $\frac{4}{5}$ of it is 20

6. 18 is $\frac{3}{4}$ of it

7. $\frac{5}{6}$ of ? = 30

8. $\frac{7}{8}$ of ? is 28

Here are some problems in the use of fractions:

9. A camera was sold for $45. The advertisement said this was $\frac{3}{4}$ of its regular price. What was the regular price?

10. After 12 gallons of gasoline were put into an empty gas tank, the needle indicated that it was $\frac{2}{3}$ full. What is the full capacity of the tank?

11. John has a 14 foot board from which he is to cut shelves for his bookcase. Each shelf is to be $2\frac{1}{3}$ ft. long. How many shelves can he get from the board?

12. At a supermarket a $5\frac{1}{2}$ lb. chicken costs $2.31. What was the cost per pound?

13. On a trip to the country the family traveled 152 miles in $3\frac{1}{6}$ hours. What was the average speed per hour?

14. How many steel screws $\frac{1}{8}$ inch apart will be needed to fasten a piece of metal $3\frac{1}{2}$ inches long?

15. One half of a birthday cake was divided among five girls. What part of the cake did each one get?

16. Martin is building a boat for resale. It will have 60 feet of moulding trim. He figures to use brass head screws every $\frac{3}{4}$ of a foot. How many screws will he need?

17. A lumber dealer knows that the plywood boards he has piled up to a height of $9\frac{3}{4}$ inches are $\frac{3}{16}$ of an inch thick. How many boards are in the pile?

18. Helen is making clothes for her doll. How many $\frac{3}{16}$ inch strips of binding can she cut out of a piece of material $\frac{3}{4}$ of an inch wide?

19. Roberta was told to rule lines $\frac{3}{8}$ of an inch apart on a sheet of blank paper. The sheet of paper was 9 inches long. How many lines would she get on the page?

20. A scout troop hiked $12\frac{3}{4}$ miles in $4\frac{1}{4}$ hours. What was the average distance traveled each hour?

# CHAPTER SIX

# LEARN TO USE DECIMALS WITH EASE

We have learned how proper fractions may be used to designate parts of a whole.

Our number system contains another method of indicating parts of a whole. The system is known as **decimal fractions** or decimals for short because it is related in every case to the number 10. It has been previously explained that the word decimal comes from the Latin word *decem* meaning ten.

Thus **decimal fractions** are fractions with denominators of 10, 100, 1000, 10,000, etc. However, the denominators are not written but rather indicated in a very ingenious way by the position of a *dot* called the *decimal point*.

To obtain a good foundation in understanding decimals, we should review our previous discussion of place values of numbers.

We have learned that numbers in the Hindu-Arabic number system have a value dependent upon their placement with respect to the decimal point. This is illustrated below.

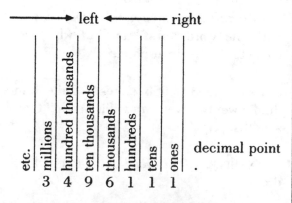

A number which appears in the second column from the right is known as tens; a number which appears in the third column

from the right is known as hundreds and so on.

In each column, **going from the right to the left,** a number was ten times as large as the same number in the column to its right. For instance, the 1 in the ten's column is as large as 10 ones. The 1 in the hundreds' column is equal to 10 tens.

## PLACE VALUES IN DECIMALS

Our system of decimal fractions is based upon the idea of place values. The decimal point becomes the dividing line between numbers greater than one and numbers having a value of less than one.

With decimals, it is still true that a number in the ones' column is ten times as large as the same number in the column *to the right of it*. A number in this column has one tenth ($\frac{1}{10}$) of the value it would have in the one's column. Similarly, a number in the next column to the right of the *tenths'* column, must have one hundredth ($\frac{1}{100}$) of the value it would have in the *ones'* column and have $\frac{1}{10}$ of the value it would have in the *tenths'* column.

Whereas previously, when dealing with whole numbers, we did not write the decimal point even though it was always considered to be present, we now must use the decimal point to separate whole numbers from numbers having a value of less than one.

We might better illustrate our place-value columns with values on both sides of the decimal point. This is shown on page 56.

The proper use of this decimal system gives us another method of working with

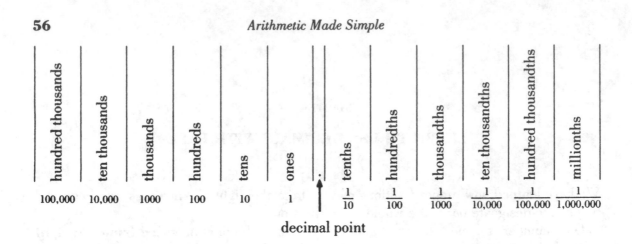

| hundred thousands | ten thousands | thousands | hundreds | tens | ones | | tenths | hundredths | thousandths | ten thousandths | hundred thousandths | millionths |
|---|---|---|---|---|---|---|---|---|---|---|---|---|
| 100,000 | 10,000 | 1000 | 100 | 10 | 1 | | $\frac{1}{10}$ | $\frac{1}{100}$ | $\frac{1}{1000}$ | $\frac{1}{10,000}$ | $\frac{1}{100,000}$ | $\frac{1}{1,000,000}$ |

decimal point

fractions in which denominators of 10, 100, 1000 or 10,000 are conveniently indicated by the placement of the decimal point. It is important to recognize that we do not write the denominator. We write only the number and put the decimal point in its proper place to denote the denominator.

## WRITING DECIMALS

*Decimals* give us a special way of writing proper fractions that have denominators ending with zero (0), or the "cipher" as it is called. For example, to write the fraction $\frac{3}{10}$ as a decimal, we write .3 because the *first place* to the right of the decimal point is the *tenths'* column.

In the same way we would write $\frac{45}{100}$ as .45 because *hundredths* are in the *second decimal place*.

In many types of machine work, where parts are shaped and ground to what is called very close tolerances, shop workers deal with measurments as small as $\frac{1}{1000}$ and $\frac{1}{10000}$ of an inch. Such fractions are generally written in decimal form as .001 and .0001 respectively. The converstion of such fractions to decimals is shown here in a general way.

When written as decimals:

| $\frac{1}{10}$ | $\frac{2}{100}$ | $\frac{3}{1000}$ | $\frac{4}{10,000}$ | $\frac{5}{100,000}$ | $\frac{6}{1,000,000}$ |
|---|---|---|---|---|---|

become
.1, .02, .003, .0004, .00005, .000006

## READING DECIMALS

In reading decimals the unit read is based on the column in which we find the *last* digit of the decimal number. As an example, .52 is read as fifty-two hundredths $\left(\frac{52}{100}\right)$. The number .502 is read five hundred two thousandths just as .052 would be read fifty-two thousandths because the last digit of these decimals is in the *thousandths'* column.

As a special case, decimals ending in zero can be read as described above or reduced. For example .350 is read as three hundred fifty thousandths $\left(\frac{350}{1000}\right)$ because the last number is in the thousandth's column even though it is a zero and could be dropped if you so desired. Actually .350 is the same as .35. They are equivalent decimals. Can you tell why? Prove it by writing in the denominators and reducing both fractions to lowest terms.

In *reading a whole number with a decimal*, the word **"and"** is read when you come to the decimal point.

EXAMPLE: 327.058 is read, three hundred twenty-seven *and* fifty-eight thousandths.

EXAMPLE: 1.01 is read one *and* one hundredth.

The steps for reading a mixed decimal consisting of a whole number and a decimal are as follows:

(a) Read the whole number as usual.

(b) Read the decimal point as "and."

(c) Then read the decimal part, naming it according to the place-value column of the last digit.

The table which follows will help you learn to read decimals of different denominations.

| PLACE OF DIGIT | HOW TO READ IT | EXAMPLE |
|---|---|---|
| First decimal place | Three tenths | $.3$ is $\dfrac{3}{10}$ |
| Second decimal place | Three hundredths | $.03$ is $\dfrac{3}{100}$ |
| Third decimal place | Three thousandths | $.003$ is $\dfrac{3}{1000}$ |
| Fourth decimal place | Three ten thousandths | $.0003$ is $\dfrac{3}{10,000}$ |
| Fifth decimal place | Three hundred thousandths | $.00003$ is $\dfrac{3}{100,000}$ |
| Sixth decimal place | Three millionths | $.000003$ is $\dfrac{3}{1,000,000}$ |

### Practice Exercise No. 42

In the parentheses ( ) next to the numbers written out in Column A, place the letter of the corresponding decimal number in Column B.

| Column A | | Column B |
|---|---|---|
| **1.** ( ) five and seven tenths | | **a** .3 |
| **2.** ( ) fifteen thousand six and nine tenths | | **b** .25 |
| **3.** ( ) three tenths | | **c** .427 |
| **4.** ( ) two and eight thousandths | | **d** .09 |
| **5.** ( ) seventy-two and sixty-three thousandths | | **e** .50 |
| **6.** ( ) twenty-five hundredths | | **f** 5.7 |
| **7.** ( ) four hundred ninety-five and sixty-seven hundredths | | **g** 2.008 |
| | | **h** 15.329 |
| **8.** ( ) four hundred twenty-seven thousands | | **i** 72.063 |
| | | **j** .0008 |
| **9.** ( ) one hundred twenty-five and seven tenths | | **k** 125.7 |
| | | **l** 24.64 |
| **10.** ( ) fifty hundredths | | **m** 285.085 |
| | | **n** 15,006.9 |
| | | **o** 495.67 |

**11.** ( ) two hundred eighty-five and eighty-five thousandths

**12.** ( ) nine hundredths

**13.** ( ) fifteen and three hundred twenty-nine thousandths

**14.** ( ) eight ten thousandths

**15.** ( ) twenty-four and sixty-four hundredths

### Practice Exercise No. 43

Write the following numbers in decimal form.

**1.** five and seven tenths

**2.** fifteen and twenty-eight hundredths

**3.** forty-two and six thousandths

**4.** two hundred twenty-three and three tenths

**5.** nine thousand twenty-nine and fifteen hundredths

**6.** $\frac{2}{10}$

**7.** $\frac{5}{1000}$

**8.** $4\frac{4}{100}$

**9.** $\frac{52}{1000}$

**10.** $\frac{32}{100}$

## ANOTHER WAY OF READING DECIMALS

It is customary in business to read large numbers like we read telephone numbers, starting at the left and naming the digits in order.

EXAMPLE 1: 3.1416 is read as:
three-point-one-four-one-six.

EXAMPLE 2: 204.713 is read as:
two-0-four-point-seven-one-three.

At times when numbers are being read to another person who is copying them, the *whole numbers* are read as usual, while in the decimal portion, the digits are read as above.

EXAMPLE 3: 2425.625 is read as:
two thousand four hundred twenty-five-point-six-two-five.

Practice reading these numbers aloud as they might be read in an office, as described above.

| | | |
|---|---|---|
| **1.** 5.602 | **5.** 150.193 | **9.** 9421.31 |
| **2.** 80.3 | **6.** 6412.70 | **10.** 2.0073 |
| **3.** 221.07 | **7.** 82,420.37 | |
| **4.** 45.006 | **8.** 78.03245 | |

## COMPARING THE VALUES OF DECIMALS

In comparing the value of proper fractions it may be necessary to find a common denominator to decide which fraction is larger.

Which is larger, $\frac{2}{3}$ or $\frac{13}{18}$? At first glance you might not know the answer.

Change $\frac{2}{3}$ to $\frac{12}{18}$. It is then obvious that $\frac{13}{18}$ is larger.

When you become more familiar with the use of decimal fractions, you will be able to tell at a glance which is the larger of two decimals. It is much easier to find the common denominators in decimal fractions than it is in common fractions.

EXAMPLE: Which is larger, .2 or .06? We know that .2 is the same as .20 or $\frac{20}{100}$ whereas .06 is $\frac{6}{100}$. Obviously .2 is larger than .06.

EXAMPLE: Which is larger, .058 or .23? Starting with the decimal point, .23 has 2 tenths and 3 hundredths, .058 has no tenths. Then .23 must be larger.

EXAMPLE: Which is larger, .734 or .62000? Starting with the decimal point, .734 has 7 in the tenths place or at least $\frac{7}{10}$. .62000 has only 6 in the tenths place or at most $\frac{6}{10}$.

Therefore, .734 must be larger.

Look at the number .62000 above. Is it larger than .62? Are they equal in value?

Keep in mind:

(a) Writing zeros *at the right-hand end* of a decimal *does not* change its value.

(b) Inserting zeros *between a decimal point and a number does change* its value.

.2 does not equal .02 or .002.

2 does not equal 20 or 200.

### Practice Exercise No. 44

Find the largest number in each group of three.

| | | | | | | |
|---|---|---|---|---|---|---|
| **1.** .3 | .5 | .40 | **6.** .8 | .79 | 1.1 |
| **2.** .4 | .42 | .04 | **7.** 53.0 | 53.001 | 53 |
| **3.** 5 | .5 | .05 | **8.** .04 | .0401 | .0048 |
| **4.** .003 | .03 | .3 | **9.** 2.91 | 2.902 | .29 |
| **5.** 1.7 | 3.01 | 5.6 | **10.** .008 | .0008 | .0079 |

## INTERCHANGE OF DECIMALS AND COMMON FRACTIONS

Since decimals are another form of fractions, it is often helpful to change from one form to the other. There are simplified methods for learning to do this easily.

**Rule: To change a decimal to a common fraction,** *remove the decimal point and write in the indicated denominator. Reduce to lowest terms.*

EXAMPLE 1: Change .25 to a common fraction.

.25 is twenty-five *hundredths* or $\frac{25}{100} = \frac{1}{4}$ ANS.

EXAMPLE 2: Change .008 to a common fraction.

.008 is eight *thousandths* or

$$\frac{8}{1000} = \frac{2}{250} = \frac{1}{125} \text{ ANS.}$$

Two methods may be used for changing a common fraction to a decimal.

EXAMPLE 1: Change $\frac{3}{4}$ to a decimal.

$$\frac{3}{4} \times \frac{25}{25} = \frac{75}{100} = .75 \text{ ANS.}$$

*Step 1.* Raise the fraction to one with a denominator of 10, 100, 1000 etc.
*Step 2.* Rewrite as a decimal.

EXAMPLE 2: Change $\frac{3}{8}$ to a decimal.

$$\frac{3}{8} = 8\overline{)3.000} \quad .375 \text{ ANS.}$$
$$\frac{2\,4xx}{60}$$
$$\frac{56}{40}$$

*Step 1.* Divide the numerator by the denominator.
*Step 2.* Write the quotient in decimal form.

**Rule: To change a proper fraction to a decimal,** *divide the numerator by the denominator.*

NOTE: *We will take up this kind of division in detail at the end of this chapter. At that time you will have a chance to practice this method with more difficult but important fractions.*

Some of the common fractions are used so often and are so closely related to our handling of money in every day use, that it is worth while memorizing them.

| Common fraction | $\frac{1}{2}$ | $\frac{1}{3}$ | $\frac{1}{4}$ | $\frac{1}{5}$ |
|---|---|---|---|---|
| Decimal equivalent | $\frac{50}{100}$ | $\frac{333}{1000}$ | $\frac{25}{100}$ | $\frac{20}{100}$ |
| Decimal form | .50 | .333* | .25 | .20 |
| Common fraction | $\frac{1}{8}$ | $\frac{3}{8}$ | $\frac{2}{3}$ | $\frac{3}{4}$ |
| Decimal equivalent | $\frac{125}{1000}$ | $\frac{375}{1000}$ | $\frac{666}{1000}$ | $\frac{75}{100}$ |
| Decimal form | .125 | .375 | .666* | .75 |

*The fractions $\frac{1}{3}$ and $\frac{2}{3}$ are distinct. Although they are used as frequently as the others they do not divide evenly when changed into decimal form. Often $\frac{1}{3}$ is written as $33\frac{1}{3}$ because of this fact. In the same way, $\frac{2}{3}$ is written $66\frac{2}{3}$ or .666.

Observe that if you know the decimal value of $\frac{1}{8}$, you can find the value for $\frac{3}{8}$ by multiplying by 3. Similarly the value of $\frac{2}{5}$ is twice that of $\frac{1}{5}$.

**Practice Exercise No. 45**

Change these decimal fractions to common fractions and reduce your answers to lowest terms.

**1.** .4  **3.** .16  **5.** .700  **7.** .1025  **9.** .8010

**2.** .05  **4.** .280  **6.** .004  **8.** .00002  **10.** .90009

Find the decimal equivalents of these common fractions to the nearest thousandth.

**11.** $\frac{3}{5}$  **13.** $\frac{5}{16}$  **15.** $\frac{1}{6}$  **17.** $\frac{5}{6}$  **19.** $\frac{9}{16}$

**12.** $\frac{5}{8}$  **14.** $\frac{12}{16}$  **16.** $\frac{4}{5}$  **18.** $\frac{7}{8}$  **20.** $\frac{9}{32}$

## HOW DECIMALS ARE RELATED TO MONEY

The monetary system of the United States is a decimal system. However, in most instances, we use only two decimal places when dealing with money numbers.

The basic unit of our money is one cent. How many cents or pennies in one dollar? One cent is therefore one hundredth, $\frac{1}{100}$ or .01 of a dollar. Now you can understand why one cent is written as $.01.

The decimal point separates the whole dollars from the cents or fractional parts of a **whole** dollar. $7.52 denotes 7 whole dollars, 5 ten-cent pieces and 2 one-penny coins. Note how these follow our place-value columns. For the decimal part we could say 5 tenths of __?__ and 2 hundredths of __?__ . To the left of the decimal we would say 7 ones. By this we would mean 7 one-dollar bills.

Let us illustrate by place values $435.62. We have 4 hundred-dollar bills, 3 ten-dollar bills, 5 one-dollar bills **and 6** tenths of a dollar plus 2 hundredths of a dollar.

Describe in place values $6758.04 __?__ thousands __?__ hundreds __?__ tens __?__ ones **and** __?__ tenths plus __?__ hundredths.

### Practice Exercise No. 46

In the exercise below assume that you are given change equal to the following amounts in the highest exact denominations of bills and coins. You are to indicate the number of coins or bills you would receive depending upon the specific question.

EXAMPLES:

| | | |
|---|---|---|
| $ .83 | How many pennies? | ANS. 3 |
| $ .91 | How many dimes? | ANS. 9 |
| $1.51 | How many half dollars? | ANS. 1 |

1. $.05 — How many dimes?
2. $1.13 — How many pennies?
3. $24.53 — How many one-dollar bills?
4. $6.40 — How many twenty-five-cent pieces?
5. $99.98 — How many fifty-dollar bills?
6. $242.13 — How many dimes?
7. $358.90 — How many pennies?
8. $5441.01 — How many one hundred-dollar bills?
9. $17,452.30 — How many ten thousand-dollar bills?
10. $42,658.19 — How many thousand-dollar bills?

### ADDITION OF DECIMALS

Although you had some practice in adding and subtracting decimals when you worked with money figures, your work was confined to the use of only two decimal places. In various kinds of engineering activities and machine shop work, it is often necessary to work with decimals that extend to five or six places, and it is not unusual to work with decimals extending beyond six places.

### *Reminders When Adding Decimals*

(a) Place the numbers in columns with the decimal points in a column.

(b) Keep hundredths under hundredths, tenths under tenths and so on.

(c) For whole numbers, keep ones under ones, tens under tens, hundreds under hundreds and so on.

(d) Add the columns the same as you would for regular addition of whole numbers, keeping the decimal point in the sum in the column of decimal points.

EXAMPLE: Find the sum of 3.61, 2.596, .3, 14.01 and .085.

METHOD:

| | | |
|---|---|---|
| 3.61 | | 3.610 |
| 2.596 | | 2.596 |
| .3 | *or* | .300 |
| 14.01 | | 14.010 |
| .085 | | .085 |
| 20.601 ANS. | | 20.601 ANS. |

EXPLANATION: Since we have one-place, two-place, and three-place numbers, we made the columns even by adding zeros to some of the numbers to act as *place holders* and fill the empty spaces in the column. This is not essential, it does not change the value but it does help to avoid errors.

### Practice Exercise No. 47

Arrange the decimals in columns and add.

**1.** .3 + .5 + .7 + .8

**2.** .33 + .7 + .08 + .65

**3.** 5.2 + 3 + 4.7 + .8

**4.** .53 + .96 + .55 + .84

**5.** 3.56 + 9.72 + 2.04 + 4.3

**6.** .079 + .026 + .04 + .085

Check each column by adding the other way.

| **7.** 2.165 | **8.** 25.3 | **9.** 1.4574 | **10.** 8.915 |
|---|---|---|---|
| .846 | 87.9 | 12.67 | 24.032 |
| 5.938 | 15. | 1.003 | 5.178 |
| 1.07 | 1.74 | 7. | 37.274 |

### Adding Decimals and Fractions

In adding a common fraction and a decimal there are two alternatives. (a) Change the common fraction to decimal form and add the decimals. (b) Convert the decimal to a common fraction and add the fractions.

EXAMPLE: Add $5\frac{1}{8}$ and 3.75.

METHOD: (a) Change to a decimal.

$$5\frac{1}{8} = \quad 5.125$$
$$+ \ 3.75$$
$$\overline{8.875}$$

METHOD: (b) Change to a fraction.

$$3.75 = 3\frac{75}{100} = \quad 3\frac{3}{4} = \quad 3\frac{6}{8}$$
$$+ \ 5\frac{1}{8} \quad + \ 5\frac{1}{8}$$
$$\overline{8\frac{7}{8}}$$

$$8.875 = 8\frac{7}{8} \text{ Ans.}$$

### Practice Exercise No. 48

Add, changing common fractions to decimals.

**1.** $6.5 + 9\frac{2}{3} + 6.4$

**2.** $1.9 + 7\frac{1}{2} + 2\frac{2}{3} + .93$

**3.** $2.73 + 8\frac{1}{2} + 6\frac{1}{3} + 8.074$

**4.** $15 + 23\frac{1}{2} + 5\frac{3}{9} + 27\frac{1}{4}$

Add, changing decimals to common fractions.

**5.** $.50 + \frac{1}{2} + .25 + \frac{1}{4}$

**6.** $.75 + 1.125 + 8\frac{1}{4} + 5\frac{1}{8}$

**7.** $3.60 + 4\frac{2}{5} + \frac{9}{10} + 2.4$

**8.** $.85 + 6\frac{1}{2} + .65 + 9\frac{2}{5}$

## SUBTRACTION OF DECIMALS

In the subtraction of decimals, the procedure is the same as for subtraction of whole numbers.

### Reminders When Subtracting Decimals

(a) Keep the decimal points under each other in a column.

(b) Keep tenths under tenths, hundredths under hundredths, etc.

(c) Keep whole number place values in proper columns.

(d) Make sure that the minuend has the same number of decimal places as the subtrahend—add zeros if needed.

(e) Keep the decimal point in its proper alignment in the final difference.

EXAMPLE 1: Subtract 9.278 from 18.3.

METHOD:

$$
\begin{array}{r}
18.3 \\
- \ 9.278 \\
\end{array}
\quad = \quad
\begin{array}{rl}
18.300 & \text{minuend} \\
- \ 9.278 & \text{subtrahend} \\
\hline
9.022 & \text{difference}
\end{array}
$$

EXAMPLE 2: Subtract 2.65 from 35.4043.

METHOD:

$$
\begin{array}{r}
35.4043 \\
- \ 2.65 \\
\end{array}
\quad = \quad
\begin{array}{rl}
35.4043 & \text{minuend} \\
- \ 2.6500 & \text{subtrahend} \\
\hline
32.7543 & \text{difference}
\end{array}
$$

Observe that in the subtraction of decimals as in the addition of decimals, writing zeros in the empty places helps to avoid errors.

Subtract and check by adding.

1.　　.74　　　5.　　28.7　　　9.　　19.1
　　− .42　　　　　− 6.7　　　　　− 14.724

2.　　9.4　　　6.　　3.05　　　10.　　25.318
　　− 7.8　　　　　− .12　　　　　　− 9.466

3.　　14.7　　　7.　　6.803
　　− 6.4　　　　　− 2.761

4.　　12.7　　　8.　　15.14
　　−  .9　　　　　− 9.348

## MULTIPLICATION OF DECIMALS

**Rule: To multiply decimals,** *proceed as in multiplication of whole numbers. But in the product, beginning at the right, point off as many decimal places as there are in the multiplier and in the multiplicand.*

EXAMPLE 1: Multiply 4.21 by .45.

METHOD:

4.21　　*multiplicand*—has two decimal places

× .45　　*multiplier*—has two decimal places
2105
1684
1.8945　　*product*—requires four decimal places

EXPLANATION: Since there is a total of four decimal places in the multiplicand and multiplier, start at the right and count four places. The decimal point then belongs to the left of the 8 which is the fourth place.

ESTIMATION OF ANSWERS: To determine whether your multiplication is *reasonably* correct, it is advisable to estimate the answer.

In the above example 4.21 × .45 you could consider that .45 is close to .5 and can be taken as $\frac{1}{2}$. Then $\frac{1}{2}$ of 4 would be about 2. The actual answer is 1.89 which is reasonably close to 2.

EXAMPLE 2: Multiply 7.33 by 3.

METHOD: Estimate 7 × 3.
　　Approximate answer is 21.

7.33　　two decimal places
× 3　　no decimal places
21.99　　*product*—requires two decimal places

EXAMPLE 3: Multiply .31 by .2

METHOD:

Estimate $\frac{1}{5}$ of $\frac{3}{10} = \frac{3}{50}$

or $\frac{6}{100}$ or .06

.31　　two places
× .2　　one place
.062

Three places are needed in the answer. Therefore we place a zero to the left of the 6 to give the correct number of decimal places.

**Rule: If there are not enough places in a product,** *put zeros to the immediate right of the decimal point to give the proper number of decimal places.*

EXAMPLE 4: Multiply .212 by .203

Estimate $\frac{1}{5} \times \frac{1}{5} = \frac{1}{25}$

or $\frac{4}{100}$ or .04

.212　　three places
× .203　　three places
636
4240
?43036
.043036

Six places are needed in the answer.

Since there are only five digits, we place a zero to the left of the 4 to give the required number of decimal places.

## Practice Exercise No. 50

Find the position of the decimal point in each product.

**1.** .15 × .32 = 480  **6.** 3.5 × .46 = 1610

**2.** .23 × 3.4 = 782  **7.** .1 × .047 = 470

**3.** .04 × 3.44 = 1376  **8.** 6.02 × 25 = 15050

**4.** 2.6 × .26 = 676  **9.** .15 × 8.7 = 1305

**5.** 6.3 × 1.33 = 8379  **10.** 53 × .34 = 1802

Estimate each answer, then multiply to get the correct answer.

**11.** 8.2 × 6.5 =  **16.** 15.25 × 36 =

**12.** .03 × .12 =  **17.** 250 × 1.8 =

**13.** .83 × .94 =  **18.** 1.75 × 51 =

**14.** 75 × .038 =  **19.** .78 × 4.8 =

**15.** 17.6 × 120.2 =  **20.** 223 × .031 =

### Multiplying Decimals by 10, 100, 1000

In multiplication of whole numbers, you learned that to multiply a number by 10, 100 or 1000 you need only add to the right of the number as many zeros as there were in the multiplier.

You remember that 453 × 10 is 4530 and 852 × 100 is 85200.

Remembering where the decimal point is in a whole number, even though it is not written, you will see that *in multiplying by 10, we moved the decimal point one place to the right*. In multiplying a whole number by 100 we moved the decimal point two places to the right.

Now observe the process with decimal numbers.

EXAMPLE 1: Multiply a decimal by 10.

METHOD: 86.35 × 10 = 863.5.
The decimal point is moved to the right one place.

EXAMPLE 2: Multiply a decimal by 100.

METHOD: 86.35 × 100 = 8635.
The decimal point is moved to the right two places.

EXAMPLE 3: Multiply a decimal by 1000.

METHOD: 86.35 × 1000 = 86350.
The decimal point is moved to the right three places. Observe that it was necessary to add a zero before the decimal point in order to provide the third place.

EXAMPLE 4: Multiply .086 by 100.

METHOD: .086 × 100 = 8.6. Moving the decimal two places to the right, we drop the zero since it is meaningless before a whole number.

If we now change the wording of the rule for multiplying whole numbers ending in zero, it will apply to all numbers including decimals.

**Rule: To multiply by 10, 100, 1000, etc.,** *move the decimal point in the multiplicand one place to the right for each zero in the multiplier.*

### Practice Exercise No. 51

Multiply the short way by moving the decimal point.

**1.** .04 × 10 =  **9.** 3.1416 × 10 =

**2.** 5.37 × 100 =  **10.** 850 × 10 =

**3.** 6.3 × 100 =  **11.** .0051 × 100 =

**4.** 852.1 × 10 =  **12.** 53 × 1000 =

**5.** .037 × 100 =  **13.** .040 × 100 =

**6.** 8.03 × 1000 =  **14.** .003 × 10 =

**7.** 16.45 × 100 =  **15.** $16.47 × 10 =

**8.** 6.137 × 100 =

### Division of Numbers by 10, 100, 1000

If it is possible to **multiply** by 10, 100, and 1000 by moving the decimal point **to the**

**right,** it should be possible to **divide** by these numbers by moving the decimal point **to the left.** Let's see how it works.

EXAMPLE 1: Divide 27.61 by 10.
Moving decimal point one place to the left.

$$27.61 \div 10 = 2.761$$

Carrying out division:

$$
\begin{array}{r}
2.761 \quad \text{ANS.} \\
10\overline{)27.610} \\
20 \text{ xxx} \\
\overline{76} \\
70 \\
\overline{61} \\
60 \\
\overline{10}
\end{array}
$$

Which method is easier?

EXAMPLE 2: .35 ÷ 10.
Moving decimal point one place to the left.

$$.35 \div 10 = .035.$$

EXAMPLE 3: 7.23 ÷ 100.
Moving decimal point two places to the left.

$$7.23 \div 100 = .0723.$$

EXAMPLE 4: 4300 ÷ 1000.
Moving decimal point three places to the left.

$$4300 \div 1000 = 4.3.$$

**Practice Exercise No. 52**

Divide the short way by moving the decimal point.

1. .87 ÷ 10 =
2. .085 ÷ 10 =
3. 297.3 ÷ 10 =
4. 387 ÷ 100 =
5. $25 ÷ 100 =
6. 7.8 ÷ 100 =
7. .063 ÷ 1000 =
8. 9.4 ÷ 10 =
9. $250 ÷ 1000 =
10. 3.87 ÷ 1000 =
11. 53.2 ÷ 100 =
12. .097 ÷ 100 =
13. $24.50 ÷ 10 =
14. 5 ÷ 10 =
15. $125 ÷ 100 =

## A Short Way to Multiply by .1, .01, .001

In studying fractions, you learned that multiplying by a fraction gave the same result as dividing by the *reciprocal*.

For example, multiplying 28 by $\frac{1}{2}$

$$\overset{14}{\cancel{28}} \times \frac{1}{\underset{1}{\cancel{2}}} = 14. \text{ The reciprocal of } \frac{1}{2} \text{ is 2.}$$

$$2\overline{)28} = 14$$

EXAMPLE 1: Multiply 28 by $\frac{1}{10}$

$$28 \times \frac{1}{10} = \frac{28}{10} = 2\frac{8}{10} \text{ or divide by the reciprocal.}$$

$$
\begin{array}{r}
2\frac{8}{10} \quad \text{ANS.} \\
10\overline{)28} \\
20 \\
\overline{8}
\end{array}
$$

Rewrite this example as a decimal. Multiply 28 by .1.

$$
\begin{array}{ll}
28 & \text{no places} \\
\times .1 & \text{one place} \\
\overline{2.8} & \text{one place in the product}
\end{array}
$$

**Rule: To multiply by .1,** *move the decimal point one place to the left (as if dividing by 10).*

EXAMPLE 2: Multiply 384 by .01. Multiply as a fraction

$$384 \times \frac{1}{100} = \frac{384}{100} = 3\frac{84}{100} \text{ ANS.}$$

Divide by the reciprocal

$$100\overline{)384} = 3\frac{84}{100}$$

Multiply as a decimal

$$
\begin{array}{rl}
384 & \text{no places} \\
\times\ .01 & \text{two places} \\
\hline
3.84 & \text{two places in the product}
\end{array}
$$

**Rule: To multiply by .01,** *move the decimal point two places to the left (as if dividing by 100).*

EXAMPLE 3: Multiply 583 by .001. Multiply as a fraction

$$583 \times \frac{1}{1000} = \frac{583}{1000} = .583 \text{ ANS.}$$

Divide by the reciprocal

$$1000\,\overline{)\,583} = .583$$

Multiply as a decimal

$$
\begin{array}{rl}
583 & \text{no places} \\
\times\ .001 & \text{three places} \\
\hline
.583 & \text{three places in the product}
\end{array}
$$

**Rule: To multiply by .001,** *move the decimal point three places to the left (as if dividing by 1000).*

### Practice Exercise No. 53

Multiply the short way by moving the decimal point.

1. .1 × 752 =
2. .01 × 12.67 =
3. .001 × 525 =
4. 10 × 1.3 =
5. 10 × .04 =
6. .01 × 7.4 =
7. .1 × .3 =
8. 10 × 293.2 =
9. .001 × 282.1 =
10. 100 × 25.9 =
11. .1 × 25.3 =
12. 10 × 2.53 =
13. 100 × 5.684 =
14. 287 × 100 =
15. .001 × .39 =

## DIVISION OF DECIMALS

In dividing with money numbers you learned to place the decimal point in the quotient directly above the decimal point in the dividend. You also learned that you must have as many places in the quotient as you have in the dividend.

As you might guess, the same principles hold true for dividing in similar decimal situations. Let us apply them to an actual problem using decimals.

### *Dividing a Decimal by a Whole Number*

PROBLEM: An automobile traveled 642.9 miles in 12 hours. What was the average rate of speed?

METHOD: Divide 642.9 by 12.

$$
\begin{array}{r}
53.57\frac{6}{12} \text{ or } 53.58 \text{ ANS.} \\
12\,\overline{)\,642.90} \\
\underline{60\text{x xx}} \\
42 \\
\underline{36} \\
69 \\
\underline{60} \\
90 \\
\underline{84} \\
6
\end{array}
$$

**Rule: To divide a decimal by a whole number,** *divide as usual, but place the decimal point in the quotient directly above the decimal point in the dividend.*

Notice that if the division process stopped with the original dividend of 642.9, the answer would by 53.5 with a remainder. To carry out the answer to another decimal place (hundredths), we add a zero in the dividend and continue the division. The division can be carried out further but it is not necessary in this situation. When the remainder is equal to a fraction of $\frac{1}{2}$ or greater, it is customary to increase the last digit by one.

PROBLEM: A dealer bought a supply of 24 electric clocks for $122.95. How much did each clock cost?

METHOD: Divide $122.95 by 24.

$$24 ) \overline{\begin{array}{l} \$\quad 5.12\tfrac{7}{24} \text{ ANS.} \\ \$122.95 \\ \underline{120 \text{ xx}} \\ \quad 29 \\ \quad \underline{24} \\ \quad 55 \\ \quad \underline{48} \\ \quad \ 7 \end{array}}$$

When the remainder is a fraction which is less than $\tfrac{1}{2}$ of the divisor it is dropped. Since the remainder in this case is $\tfrac{7}{24}$ we drop the 7 as it is less than $\tfrac{1}{2}$ of 24.

### Dividing a Decimal by a Decimal

PROBLEM 1: How many aluminum washers .9 inch thick can be sliced from a piece of aluminum tubing 13.5 inches long?

METHOD: Divide 13.5 by .9.

$$.9 ) \overline{\begin{array}{l} 1\ 5. \\ 13.5. \\ \underline{9 \text{ x}} \\ 4\ 5 \\ \underline{4\ 5} \\ 0 \end{array}}$$

**Rule: To divide a decimal by a decimal,** *move the decimal point of the divisor as many places to the right as are necessary to make it a whole number. Next move the decimal point of the dividend the same number of places, adding zeros if necessary.*

Observe that moving the decimal point one place to the right is the same as multiplying by 10. You will recall that if we multiply the divisor and dividend by the same number, it does not change the quotient.

PROBLEM 2: The price of admission to an outdoor movie was set at $1.30 a car. The total receipts for the evening were $551.20. How many cars were admitted?

METHOD: Divide $551.20 by $1.30.

$$1.30 ) \overline{\begin{array}{l} 4\ 24. \\ 551.20. \\ \underline{520 \text{ xx}} \\ 31\ 2 \\ \underline{26\ 0} \\ 5\ 20 \\ \underline{5\ 20} \\ 0 \end{array}}$$

(a) Move decimal point two places in divisor to make it a whole number.

(b) Move decimal point the same number of places in the dividend.

(c) Put decimal point in quotient directly above its new place in the dividend.

(d) Divide as usual.

Check your answers by multiplying the quotient by the divisor.

### Practice Exercise No. 54

In the examples which follow, divide until there is no remainder. Estimate your answers first.

1. $14 ) \overline{19.6}$
2. $210 ) \overline{4.41}$
3. $3.5 ) \overline{1225}$
4. $60 ) \overline{1.5972}$
5. $8 ) \overline{16.24}$
6. $2.4 ) \overline{372}$
7. $2.3 ) \overline{62.376}$
8. $.81 ) \overline{46.413}$
9. $1.6 ) \overline{117.92}$
10. $.9 ) \overline{7.155}$

### Rounding Decimals

At times, in dividing with decimals there is a remainder. Zeros may be added to the dividend and more decimal places found. However, some quotients will never come out even. For example, try dividing one by three. It is, therefore, customary to tell the student how many decimal places are needed or desired in the quotient of any division example.

The instructions may be to find the answer to the **nearest tenth.** To do this, carry out the division to **two decimal** places. If the digit in the hundredth's place is five or more, increase the number in the tenth's place by one.

EXAMPLE: Round to the nearest tenth.
Original quotient

> 25.63 becomes 25.6
> 87.35 becomes 87.4
> 1.09 becomes  1.1

If you are instructed to find the answer to the **nearest hundredth,** carry out the division to **three decimal places.** If the digit in the thousandth's place is five or more, increase the number in the hundredth's place by one.

EXAMPLE: Round to the nearest hundredth.
Original quotient

> 247.541 becomes 247.54
> 27.085 becomes  27.09
> 129.326 becomes 129.33

### Practice Exercise No. 55

The following exercise will test your ability to round off decimals. Round off to the nearest tenth.

**1.** 21.46　**4.** 18.08　**7.** 3.12　**10.** .41

**2.** 5.83　**5.** 102.39　**8.** 19.74

**3.** 6.67　**6.** 24.76　**9.** 9.98

Round off to the nearest hundredth.

**11.** 2.624　**14.** 4.328　**17.** 3.275　**20.** 16.324

**12.** 6.071　**15.** 1.096　**18.** 25.666

**13.** 20.015　**16.** 11.255　**19.** 102.285

### Practice Exercise No. 56

Estimate the answer first. Find the quotient to the nearest tenth.

**1.** $12.25\overline{)471.68}$　　**6.** $6.23\overline{)85.73}$

**2.** $8.5\overline{)23.53}$　　**7.** $7.21\overline{)15.97}$

**3.** $5.4\overline{)52.16}$　　**8.** $64.2\overline{)783.29}$

**4.** $1.8\overline{)4.007}$　　**9.** $7.1\overline{)46.24}$

**5.** $4.6\overline{)23.745}$　　**10.** $.86\overline{)48.203}$

Find the quotient to the nearest hundredth.

**11.** $.35\overline{)48.38}$　　**16.** $4.5\overline{)6.7943}$

**12.** $3.14\overline{)457.3}$　　**17.** $7.3\overline{)84.879}$

**13.** $.28\overline{)2876}$　　**18.** $9.07\overline{)35.748}$

**14.** $.35\overline{)3.1314}$　　**19.** $750\overline{)6.913}$

**15.** $7.7\overline{)63.466}$　　**20.** $6.7\overline{)852.714}$

## FRACTION AND DECIMAL RELATIONSHIPS

Earlier in this chapter you learned how to change decimals to fractions and vice versa. You were told that you would have an opportunity to practice changing important but difficult fractions to decimals.

REMEMBER: To change a fraction to a decimal, divide the numerator by the denominator. For example,

$$\frac{1}{6} = 6\overline{)1.0000} \quad \begin{array}{c} .1666 \quad \text{etc.} \\ \text{etc.} \end{array}$$

This type of division is carried out to three or four places or more depending upon the need. Where we are using instruments that measure as closely as a ten-thousandth of an inch, our arithmetic has to be equally accurate. We would then use decimals to at least four places. If we can only measure up to thousandths, then our arithmetic work in decimals is carried out to the nearest thousandth.

An ordinary ruler is often divided into as many as 64 parts to the inch. Because it is easier to figure with 10, 100 and 1000, many rulers are divided into 10ths instead of 16ths or 32nds. For general use, however, you will find your ruler divided into fractional parts of 16ths, 32nds, or 64ths. In many kinds of daily situations you will find yourself going from fractions to decimal equivalents and back again. Automatically, you think of $\frac{1}{4}$ of a dollar as $.25. The employee in a sheet metal shop never talks of metal as $\frac{1}{16}$ inch thick, he says .0625.

**Practice Exercise No. 57**

Find the decimal equivalents for each fraction below and arrange them in the Table of Decimal Equivalents. Carry your work to four places.

### Table of Decimal Equivalents

(a) $\frac{1}{8} =$    $\frac{1}{2} =$    $\frac{3}{4} =$

$\frac{1}{4} =$    $\frac{5}{8} =$    $\frac{7}{8} =$

$\frac{3}{8} =$

(b) $\frac{1}{16} =$    $\frac{7}{16} =$    $\frac{13}{16} =$

$\frac{3}{16} =$    $\frac{9}{16} =$    $\frac{15}{16} =$

$\frac{5}{16} =$    $\frac{11}{16} =$

(c) $\frac{1}{32} =$    $\frac{13}{32} =$    $\frac{25}{32} =$

$\frac{3}{32} =$    $\frac{15}{32} =$    $\frac{27}{32} =$

$\frac{5}{32} =$    $\frac{17}{32} =$    $\frac{29}{32} =$

$\frac{7}{32} =$    $\frac{19}{32} =$    $\frac{31}{32} =$

$\frac{9}{32} =$    $\frac{21}{32} =$

$\frac{11}{32} =$    $\frac{23}{32} =$

(d) $\frac{1}{64} =$    $\frac{23}{64} =$    $\frac{45}{64} =$

$\frac{3}{64} =$    $\frac{25}{64} =$    $\frac{47}{64} =$

$\frac{5}{64} =$    $\frac{27}{64} =$    $\frac{49}{64} =$

$\frac{7}{64} =$    $\frac{29}{64} =$    $\frac{51}{64} =$

$\frac{9}{64} =$    $\frac{31}{64} =$    $\frac{53}{64} =$

$\frac{11}{64} =$    $\frac{33}{64} =$    $\frac{55}{64} =$

$\frac{13}{64} =$    $\frac{35}{64} =$    $\frac{57}{64} =$

$\frac{15}{64} =$    $\frac{37}{64} =$    $\frac{59}{64} =$

$\frac{17}{64} =$    $\frac{39}{64} =$    $\frac{61}{64} =$

$\frac{19}{64} =$    $\frac{41}{64} =$    $\frac{63}{64} =$

$\frac{21}{64} =$    $\frac{43}{64} =$

**Practice Exercise No. 58**

Problems in the use of decimals.

**1.** If you spent $395.40 for food during the month of June, what was the average cost of food each day? (Do you know how many days there are in June?)

**2.** On a vacation trip, Sam's father used 9.4 gallons of gasoline. The car travels 14.3 miles on a gallon. How far did they travel?

**3.** In the shop there are 32 sheets of copper piled on top of each other. What is the height of the pile if each sheet is .087 inches thick?

**4.** At a bazaar, Janice bought a piece of cloth that was $6\frac{1}{4}$ yards in length. She paid $12.15 for the piece. What did it cost per yard? (To the nearest cent.)

**5.** How far will an airplane go in 3.6 hours, traveling 358.4 miles per hour?

**6.** An airplane can carry 5,425.6 gallons of fuel. Gasoline weighs 5.8 lb. per gallon. What is the weight of the fuel when the tanks are full?

**7.** Peter went on a trip by bicycle. He traveled 8.9 miles in 2.4 hours. How many miles did he travel in one hour? (To the nearest tenth.)

**8.** A cylindrical storage bin containing a thousand bags of potatoes is filled to a height of 10.9 feet. How many bags would there be at a height of one foot? (To the nearest whole number.)

**9.** A stack of thin gage aluminum sheets of .0156 inch thickness make a pile 1.5 inches high. How many sheets are there in the pile? (To nearest whole number.)

**10.** Traveling at 30 miles per hour on a trip that measured 457.6 miles, Mr. Spivak averaged 20.8 miles per gallon. If the *rate* of his gas consumption is increased by $\frac{1}{2}$ when he travels at 60 miles per hour, how many gallons of gas would he use on the 457.6 mile trip at the 60 mile per hour speed?

# CHAPTER SEVEN

# PERCENTAGE

| R | R |   | R | R | R | R | R | R | R |
|---|---|---|---|---|---|---|---|---|---|
| R | R | R | R |   | R | R | R | R | R |
| R |   | R | R |   | R |   | R | R | R |
|   | R |   | R |   | R | R | R |   | R |
|   | R | R | R |   |   | R | R | R | R |
| R |   | R | R | R | R | R | R | R | R |
|   |   | R | R | R |   |   | R | R | R |
| R | R |   | R | R | R | R |   |   | R |
| R |   | R | R | R |   | R | R | R | R |
| R | R | R | R | R | R | R | R | R | R |

The square above contains 100 boxes which represent a total of 100 predictions of rain by a local weather bureau. When rain was correctly predicted, an R was marked in the box. The square contains 80 boxes with R's in them.

The weather bureau's prediction score can be expressed in any one of four ways as indicated below.

(a) 80 out of 100

(b) $\frac{80}{100}$ of the predictions

(c) $\frac{4}{5}$ of the predictions

(d) 80 percent correct

**Percent** *tells how many out of 100.*

Percentage then is a way of expressing fractional parts of 100 in arithmetic. The word *percent* means *hundredths*. Instead of writing $\frac{80}{100}$ we write 80 percent or 80%. This is the percent sign (%).

Another way of defining percent is to say: *Percent is a fraction with a denominator of 100 in which the sign % is substituted for the denominator.*

Since all figures with the % sign have a common denominator—100—percents may be added, subtracted, multiplied or divided just as other specific denominations are treated.

Thus:
$$7\% + 8\% = 15\%$$
$$22\% - 7\% = 15\%$$
$$9\% \times 3\% = 27\%$$
$$16\% \div 4\% = 4$$

In doing the actual arithmetic in a problem with percents, the % sign is not used. The percentages must be changed to a common fraction or a decimal before carrying out the operations.

**Rule: To change a percent to a common fraction or mixed number,** *drop the percent sign and write it as a fraction with 100 in the denominator. Reduce to lowest terms.*

EXAMPLE 1: Change 55% to a fraction.

METHOD: $55\% = \dfrac{55}{100} = \dfrac{11}{20}$

EXAMPLE 2: Change $7\frac{1}{2}\%$ to a fraction.

METHOD: $7\frac{1}{2} = \dfrac{\frac{15}{2}}{100} = \dfrac{\overset{3}{\cancel{15}}}{2} \times \dfrac{1}{\underset{20}{\cancel{100}}} = \dfrac{3}{40}$

EXAMPLE 3: Change 175% to a fraction.

METHOD: $\dfrac{175}{100} = 1\frac{3}{4}$

**Practice Exercise No. 59**

Complete the table below by finding the fractional equivalents of all the percentages listed. Use the method explained above.

Fractional equivalents of percents

| | | |
|---|---|---|
| $10\% = \frac{1}{10}$ | $12\frac{1}{2}\% = \frac{1}{8}$ | $8\frac{1}{3}\% = \frac{1}{12}$ |
| $20\% =$ | $25\% =$ | $16\frac{2}{3}\% =$ |
| $30\% =$ | $37\frac{1}{2}\% =$ | $33\frac{1}{3}\% =$ |
| $40\% =$ | $62\frac{1}{2}\% =$ | $66\frac{2}{3}\% =$ |
| $50\% =$ | $87\frac{1}{2}\% = \frac{7}{8}$ | $83\frac{1}{3}\% = \frac{5}{6}$ |
| $60\% = \frac{3}{5}$ | | |

**Rule: To change a percent to a decimal,** *drop the percent sign and multiply by* $\frac{1}{100}$ *or .01.* A short way is to *move the decimal point two places to the left.*

EXAMPLE 1: Change 28% to a decimal.

$28\% = 28 \times .01 = .28$
$28\% = .28$ ANS.

EXAMPLE 2: Change 2.3% to a decimal.

$2.3\% = 2.3 \times .01 = .023$
$2.3\% = .023$ ANS.

Moving the decimal point two places to the left, meant one zero had to be added.

**Practice Exercise No. 60**

Change the percents below to decimals.

| | | |
|---|---|---|
| **1.** 3% | **6.** 20% | **11.** 65% |
| **2.** 100% | **7.** 200% | **12.** 90% |
| **3.** 5% | **8.** 6% | **13.** 53.5% |
| **4.** 25% | **9.** 40% | **14.** 95% |
| **5.** 50% | **10.** 71% | **15.** 125% |

**Rule: To change a decimal to a percent,** *multiply by 100 and add the percent sign.* In effect, *move the decimal point two places to the right.*

EXAMPLE 1: Change .32 to a percent.
$.32 \times 100 = 32 = 32\%$ ANS.

EXAMPLE 2: Change .04 to a percent.
$.04 \times 100 = 4 = 4\%$ ANS.

EXAMPLE 3: Change 2.5 to a percent.
$2.5 \times 100 = 250 = 250\%$ ANS.

Any whole number greater than one which designates a percent is more than 100%.

EXAMPLE 4: Change .0043 to a percent. $.0043 = .43\%$. Less than 1%. This was done by the short method of moving the decimal point two places to the right.

**Practice Exercise No. 61**

Change the numbers below to percents.

| | | |
|---|---|---|
| **1.** .20 | **6.** .253 | **11.** .238 |
| **2.** .06 | **7.** 2.53 | **12.** .45 |
| **3.** .10 | **8.** .125 | **13.** .6 |
| **4.** .4 | **9.** 1.00 | **14.** .01 |
| **5.** .47 | **10.** .02 | **15.** .5 |

**Rule: To change a fraction to a percent,** *express the fraction in hundredths, then express the hundredths as a percent.*

EXAMPLE 1: Change $\frac{1}{4}$ to a percent.

METHOD: $\frac{1}{4} = \frac{25}{100} = 25\%$ ANS.

Another procedure to changing a fraction to a percent is: *divide the numerator by the denominator, multiply the quotient by 100 and add the % sign.*

EXAMPLE 2: Change $\frac{2}{5}$ to a percent.

(a) $\frac{2}{5} \times 100 = \frac{200}{5} = 40\%$ ANS. or

(b) $5\overline{)2.0}$  $.4 \times 100 = 40\%$ ANS.

**Practice Exercise No. 62**

Change the fractions below to percents.

| | | | | |
|---|---|---|---|---|
| **1.** $\frac{1}{2} =$ | **4.** $\frac{7}{8} =$ | **7.** $\frac{1}{3} =$ | **10.** $\frac{1}{4} =$ | **13.** $\frac{2}{9} =$ |
| **2.** $\frac{1}{20} =$ | **5.** $\frac{11}{50} =$ | **8.** $\frac{1}{12} =$ | **11.** $\frac{1}{8} =$ | **14.** $\frac{2}{3} =$ |
| **3.** $\frac{3}{4} =$ | **6.** $\frac{5}{8} =$ | **9.** $\frac{1}{5} =$ | **12.** $\frac{3}{8} =$ | **15.** $\frac{3}{7} =$ |

## MEMORIZING FRACTIONAL EQUIVALENTS OF POPULAR PERCENTS

Since percents are really fractions and a form of decimals, you should have at your fingertips the equivalent forms of the most used fractions or percentages. While it is important to know how to convert percents to fractions and decimals, it is also important to learn to interchange from memory the most commonly used fractions, decimals and percents.

### Practice Exercise No. 63

Fill in the spaces in the chart and memorize it when it is completed.

*Table of Popular Percent—Fraction—Decimal Equivalents*

| Percent | Common Fraction | Decimal |
|---|---|---|
| | $\frac{1}{2}$ | |
| $33\frac{1}{3}$ | | |
| | | .25 |
| | $\frac{1}{8}$ | |
| $37\frac{1}{2}$ | | |
| | $\frac{2}{3}$ | |
| | | .75 |
| | $\frac{5}{8}$ | |
| $87\frac{1}{2}$ | | |

## COMPARING PERCENTS

You will recall that in order to compare fractions, they had to be changed so that they had common denominators.

In certain problems you will find some quantities stated as decimals and some as fractions or percents. In order to compare them, you must change them *all* to either fractions, decimals or percents. You have had experience in making these types of changes.

Test yourself on the comparisons below. You should be able to do most of them without the use of pencil and paper.

### Practice Exercise No. 64

Select the smaller value of each pair.

**1.** 20% or $\frac{1}{6}$      **6.** $62\frac{1}{2}$% or .64

**2.** $\frac{1}{5}$ or 18%      **7.** 15% or $\frac{1}{7}$

**3.** $\frac{1}{3}$ or $37\frac{1}{2}$%      **8.** 19% or .2

**4.** 75% or $\frac{5}{8}$      **9.** $\frac{7}{8}$ or 90%

**5.** 75% or $\frac{7}{8}$      **10.** $\frac{1}{8}$ or .1

Select the smallest value of each group.

**11.** $66\frac{2}{3}$%, .63, $\frac{5}{8}$      **16.** .9, $\frac{1}{10}$, $12\frac{1}{2}$%

**12.** $\frac{1}{3}$, 35%, .4      **17.** .07, $66\frac{2}{3}$%, $\frac{5}{8}$

**13.** $\frac{7}{8}$, 90%, .88      **18.** $\frac{1}{4}$, 20%, $.22\frac{1}{2}$

**14.** $\frac{1}{3}$, .4, $37\frac{1}{2}$%      **19.** 75%, $\frac{5}{8}$, $.83\frac{1}{3}$

**15.** $\frac{1}{5}$, .25, $28\frac{2}{7}$%      **20.** .9, $\frac{7}{8}$, $83\frac{1}{3}$%

Arrange each group in order from smallest to largest.

**21.** $87\frac{1}{2}$%, .9, $83\frac{1}{3}$      **24.** $\frac{1}{5}$, $12\frac{1}{2}$%, $.16\frac{2}{3}$

**22.** $\frac{5}{8}$, $66\frac{2}{3}$%, .675      **25.** $\frac{2}{3}$, 75%, .8

**23.** $12\frac{1}{2}$%, $\frac{1}{5}$, .25

## USES OF PERCENT

In learning about fractions, we discovered that fractions could be used to give three different kinds of information.

**Percentage** is just another way of talking and writing about fractions and, therefore, is used for the same purposes as fractions. Percentage is used in:

(a) Finding the size or value of part of a number or sum of money;

(b) Finding out what part one number is of another;

(c) Finding the value of a whole quantity when we know only a part.

The arithmetic procedures in this section will for the most part be a review. What you will have to learn primarily are the new terms and how to apply them properly.

### Percent of a Number

How do we find a *percent* of a number? Compare this with finding a *part* of a number.

EXAMPLE 1: How much is 40% of 540?

METHOD (a)
Using fractions

$$40\% = \frac{40}{100} = \frac{2}{5}$$

$\frac{2}{5}$ of 540 is the same as

$$\frac{2}{\cancel{5}} \times \frac{\cancel{540}^{108}}{1} = 216 \text{ ANS.}$$

METHOD (b)
Using decimals

$$40\% = .4$$

$$\begin{array}{r} 540 \\ \times\ .4 \\ \hline 216.0 \text{ ANS.} \end{array}$$

REMEMBER: In order to work with percents, you must convert them first to either fractions or decimals.

Any percent can be converted to a fraction by making the denominator 100 and removing the % sign.

Any percent can be converted to a decimal by moving the decimal point two places to the left and removing the % sign.

EXAMPLE 2: Find 37½% of 120.

METHOD (a) by fractions

$$37\tfrac{1}{2}\% = \frac{37\tfrac{1}{2}}{100} = \frac{3}{8} \text{ (see Table)}$$

$$\frac{3}{\cancel{8}} \times \frac{\cancel{120}^{15}}{1} = 45 \text{ ANS.}$$

METHOD (b) by decimals

$$37\tfrac{1}{2}\% = .375$$

$$\begin{array}{r} 120 \\ \times\ .375 \\ \hline 600 \\ 840 \\ 360 \\ \hline 45.000 \text{ ANS.} \end{array}$$

EXAMPLE 3: Find 2% of $300.

METHOD (a) by fractions

$$2\% = \frac{2}{100} = \frac{1}{50}$$

$$\frac{1}{\cancel{50}} \times \frac{\cancel{\$300}^{6}}{1} = \$6 \text{ ANS.}$$

METHOD (b) by decimals

$$2\% = .02$$

$$\begin{array}{r} \$300 \\ \times\ .02 \\ \hline \$6.00 \text{ ANS.} \end{array}$$

EXAMPLE 4: Find ½% of $500.

METHOD (a) by fractions

$$\frac{1}{2}\% = \frac{\tfrac{1}{2}}{100} = \frac{1}{2} \times \frac{1}{100} = \frac{1}{200}$$

$$\frac{1}{\cancel{200}_{2}} \times \frac{\cancel{\$500}^{5}}{1} = \frac{5}{2} = \$2\tfrac{1}{2} \text{ ANS.}$$

METHOD (b) by decimals

$$\frac{1}{2}\% = .005$$

$$\begin{array}{r} \$500 \\ \times\ .005 \\ \hline \$2.500 \text{ Ans.} \end{array}$$

You can see from the examples above that in some instances it is easier to carry out the arithmetic computations with fractions, while in others it is simpler to work with decimals. In Example 2 above, it is obviously easier to take $\frac{3}{8}$ of 120 than to multiply 120 by .375. The method used is unimportant. The answers will be the same both ways.

### Practice Exercise No. 65

Find the answers to the examples below, to the nearest hundredth, using either fractions or decimals.

1. 5% of 140
2. 20% of $30.50
3. 25% of 832
4. $33\frac{1}{3}\%$ of $90
5. $37\frac{1}{2}\%$ of 248
6. $66\frac{2}{3}\%$ of $2.40
7. 75% of 720
8. 62.5% of 176
9. $87\frac{1}{2}\%$ of 552
10. 70% of 200
11. 6% of $1200
12. 3% of $1500
13. 45% of 180
14. 125% of 120
15. 2.5% of $50

### Shortcuts for Finding Percents Mentally

Some savings institutions pay 1% for the use of your money at three-month intervals. If you had $215 in savings at such a company, what would you get for the three-month period?

1% = .01. We can multiply by .01, by *moving the decimal point* **two** *places to the left.*

EXAMPLE: 1% of $215 = $2.15.

In the same way, we find 10% by *moving the decimal point* **one** *place to the left.*

EXAMPLE: 10% of $523.50 = $52.35.

You can apply these procedures to figure other percentages mentally.

To find 11% of a decimal or money number—find 1% and 10% mentally and add them together.

EXAMPLE: 11% of $25. Take 10% = $2.50, 1% = $.25. Total = $2.75. Easier than multiplying .11 × $25.

To find a familiar fractional part of 1%—find 1% and divide.

EXAMPLE: $\frac{1}{2}\%$ of $250.

1% is $2.50. $\frac{1}{2}$ of $2.50 is $1.25. This is easier than multiplying $250 by .005.

### Practice Exercise No. 66

Do the following examples mentally. Use the shortcuts discussed above.

1. 1% of 25
2. 10% of $9.80
3. 1% of $4.70
4. 10% of $52.80
5. 5% of 53
6. 1% of 974
7. 10% of 1845
8. 1% of 850
9. 10% of $23
10. 1% of $23
11. 2% of $5
12. 5% of $25
13. $\frac{1}{3}\%$ of $9
14. 6% of $250
15. $\frac{1}{2}\%$ of $30

### Practice Exercise No. 67

Solve the problems in percentage below.

1. The Jukes bought a refrigerator for $680. They had to pay 12% down and the remainder in 24 equal installments. How much money did they pay down?
2. The Hawks of the little league played 24 games and won 75% of them. How many games did they win?

**3.** Allan's father earns $37,800 a year. His employer deducts $3\frac{1}{2}$% for the pension plan. How much is deducted?

**4.** Ipswich took a test that had 240 questions. His grade was 80%. How many did he get right?

**5.** Ellie May sold $35 worth of Christmas cards. She was allowed to keep 20% for herself. How much did she earn?

**6.** The city sales tax is 3%. How much tax must be paid on the purchase of a bicycle selling for $42?

**7.** An automobile is priced at $8720. The dealer will allow $12\frac{1}{2}$% off to promote a quick sale. How much does he allow?

**8.** Lois earns $142.50 a week. She is to receive a 10% increase at the end of a year. How much will she earn then?

**9.** To purchase traveler's checks, the banks charge a rate of 1%. What does it cost to purchase $350 in checks?

**10.** In addition to his salary Mr. Kallipak receives a commission of $\frac{3}{4}$% on the sales he makes. How much extra would he earn in a week if he sold $760 worth of merchandise?

### Finding What Percent One Number is of Another

How do we find what *percent* one number is of another? Compare this with finding what *part* one number is of another.

EXAMPLE 1: In a basketball game, the home team made 12 foul goals out of 24 tries. What percent of the attempts were successful?

METHOD:
12 is what part of 24?

$$\frac{12}{24} = \frac{1}{2}$$

$$\frac{1}{2} = 50\% \text{ ANS.}$$

*Step 1.* Write the numbers compared as a fraction.
*Step 2.* Change the fraction to a percent.
In such problems as above, there may be some difficulty in identifying the numerator and the denominator. Look for the word **of**, since it is usually associated with the whole or **denominator.**

At times, you will find it helpful to identify the **numerator** as preceding the word **is.**

EXAMPLE 2: 50 **is** what part **of** 125?

METHOD: numerator     denominator

$$\frac{50}{125} \times \frac{2}{5}$$

$$\frac{2}{5} = 40\% \text{ ANS.}$$

*Step 1.* Write the numbers compared as a fraction.
*Step 2.* Change the fraction to a percent.

An example similar to the above could be worded differently.

EXAMPLE 3: What percentage is 150 of 60?

numerator denominator

METHOD: $\frac{150}{60} = \frac{15}{6} = 2\frac{1}{2}$

$$2\frac{1}{2} = 250\% \text{ ANS.}$$

*Step 1.* Write numbers as a fraction.
*Step 2.* Change the fraction to a percent.
Notice that you should not rely on using the larger number as the denominator. In Example 3, the larger number happens to be the numerator. This will be true in any example when the percentage is greater than ? .

**Practice Exercise No. 68**

Do the examples below carefully.

**1.** What % is $1.25 of $2.50?

**2.** 15 is what % of 45?

**3.** In %, what part of 28 is 7?

**4.** What % is $24 of $84?

**5.** What % is 15 of 10?

**6.** What % is 10 of 15?

**7.** 60 is what % of 40?

**8.** 3 is what % of 100?

**9.** What percent is 175 of 150?

**10.** What percent is 6 of 200?

### Practice Exercise No. 69

Solve the problems in percentage below.

**1.** Out of 150 shots in a target shoot, John hit the bull's-eye 120 times. What percent of his shots hit the bull's-eye?

**2.** A small boat kit that regularly sells for $50 is advertised for $37.50. At what percent of the regular price is it being sold?

**3.** Sam's trip from his home to a vacation spot on Lake George is 250 miles of which 160 miles are on the New York Thruway. What percentage of the trip is on the Thruway?

**4.** Ronald got 800 eggs from his chicken brood, of which 450 were brown. What percent were brown eggs?

**5.** Mr. Keller bought a watch for $80. The clerk told him he would have to pay an additional $12 in tax. What percent of the original price of the watch was the tax?

**6.** The school baseball team won 16 games and lost 9 last season. What was their percentage of wins?

**7.** Debra earns an average of $9 per week baby sitting. She pays $3 a week for a newspaper advertisement to get customers. What percent of her earnings remains as a profit?

**8.** Frank started a newspaper route with 125 customers. At the end of three months he had only 100 customers. What percent of his customers did he lose?

**9.** There were 40 scouts present at a father and son outing, but only 35 fathers were able to attend. What percent of the fathers could not attend?

**10.** Spike won an $80 camera in a raffle. He sold it to Hock Shop Harry for $38. What percent of its price did Spike sacrifice?

### *Finding a Number When a Percent of It Is Given*

How do we find the whole when a *percent* of it is given? Compare this with finding the whole when a *part* is given.

PROBLEM: A manufacturing company employs 200 machinists. This is 40% of all their employees. What is the total number of people they employ?

METHOD (a): We are saying, 200 is $\frac{2}{5}$ of what?

$$40\% = \frac{40}{100} \text{ or } \frac{2}{5}$$

$$200 \div \frac{2}{5} =$$

$$\frac{\overset{100}{\cancel{200}}}{1} \times \frac{5}{\underset{1}{\cancel{2}}} = 500 \text{ Ans.}$$

*Step 1.* Change the percent to fraction.
*Step 2.* To find the whole when a fractional amount is given, **divide** the amount by the given fraction.

METHOD (b): 200 is 40% of ?
$$40\% = .40 \text{ or } .4$$

$$200 \div .4 \text{ or } .4\overline{)200.0}\phantom{0}^{500 \text{ Ans.}}$$

*Step 1.* Change the percent to a decimal.
*Step 2.* Divide as previously.

There is still another method for doing this type of example. It is more of a reasoning process. If you learn this method, you will be able to apply it in many more difficult types of problems later on.

METHOD (c):
**Since 200 employees is 40%, then 1% would be $\frac{1}{40}$ of 200 or 5 employees.**

**If 1% = 5, then 100% = 5 × 100 = 500 Ans.**

**In our judgment, method (c) is the easiest to remember and apply in doing such problems. Once you grasp this method, you are not apt to forget it. Methods (a) and (b) are more of a rote memory process and will be more readily forgotten over a period of time.**

### Practice Exercise No. 70

Do the examples below carefully.

**1.** 75 is 5% of ?       **6.** 80 = 8% of ?

**2.** 30 is 3% of ?       **7.** 6 = 1% of ?

**3.** $50 is 20% of ?       **8.** 12 = 200% of ?

**4.** 200 = $33\frac{1}{3}$% of ?       **9.** 150% of ? is $30

**5.** 15 is 5% of ?       **10.** 125% of ? is 125

### Practice Exercise No. 71

Solve the percentage problems below.

**1.** Gene intends to buy a used bicycle. He has saved $15, which is $37\frac{1}{2}$% of the cost. What will the bike cost?

**2.** Shorty sells magazines and is allowed to keep 10% of the money he collects. Last month he earned $21.30. How much money did he collect for the month?

**3.** A pet store reduced the price on a puppy by $9 which was 30% of its price. What is the price?

**4.** The ad stated that the salesman gets 20% of the sales price of the books he sells and that many were earning $50 part time. How many dollars worth of books must Fred sell to earn $50?

**5.** A basketball team "hit" with 27 field shots for the basket. This was 30% of the tries they had taken. How many shots had they attempted?

**6.** A special trip to Washington, D.C. was advertised at $215. This was 75% of the regular fare. What was the regular fare?

**7.** Ely bought some tropical fish. Twelve of them died in the first week. This was 8% of his total. How many did he have at the start?

**8.** The Eisensteins rented a summer cottage for $1200. If the rental cost is 5% of the value of the house, what is its value?

**9.** Bob Brady owns 40% of a hardware store. His share cost $84,000. What was the total cost of the store?

**10.** Carolyn manages to save 20% of her allowance each week. She put $19.20 in the bank at the end of 8 weeks. How much does she get for a weekly allowance?

### Rounding Off Percents

Percents may be rounded off in the same manner as decimals. Sometimes the quantity is rounded to the nearest one percent, the nearest tenth of a percent or the nearest hundredth of a percent, depending upon the accuracy needed.

EXAMPLE 1: Change $\frac{3}{14}$ to a percent and round to the nearest $\frac{1}{10}$ of a percent.

$$\text{METHOD: } 14 \overline{)3.0000} \quad \frac{.2143}{} = 21.4\% \text{ Ans.}$$

$$\frac{2\ 8}{20} \quad \text{or } 21\frac{4}{10}\% \text{ Ans.}$$

$$\frac{14}{60}$$

$$\frac{56}{40}$$

REMEMBER: To change a decimal to a percent, move the decimal point two places to the right and add the % sign.

EXAMPLE 2: Change .57167 to a percent and round to the nearest $\frac{1}{100}$ of a percent.

METHOD: .5717 = $57\frac{17}{100}$% Ans.

REMEMBER: If last digit is five or more, increase the preceding digit by one. If it is less than five, drop it.

### Practice Exercise No. 72

Round to the nearest whole percent.

**1.** 13.7%       **4.** 96.1%

**2.** 54.2%       **5.** 91.8%

**3.** 68.5%

Round to the nearest $\frac{1}{10}$ percent.

**6.** 14.3%       **9.** 56.44%

**7.** 76.8%       **10.** 61.48%

**8.** 82.25%

Round to the nearest $\frac{1}{100}$ percent.

**11.** 19.382%       **14.** 134.682%

**12.** 65.716%       **15.** 79.996%

**13.** 23.495%

## USING PERCENTS TO SHOW CHANGES

Percentage is often used to show an amount of increase or decrease in a quantity or value. The portion of such change is expressed as a percent.

Usually, problems involving percentage of changes or variations are two-step problems. They require a preliminary computation before you can attempt to find the percentage.

PROBLEM: Last year 200 boys entered the Soap Box Derby in our town. This year there were 220 entries. What was the percent of increase this year over last?

METHOD: (a) $220 - 200 = 20$

(b) $\dfrac{20}{200} = \dfrac{10}{100}$

(c) $\dfrac{10}{100} = 10\%$ ANS.

*Step 1*. Find *amount* of difference.
*Step 2*. Write a fraction of the two amounts being compared.
*Step 3*. Express the fraction as a percent.

PROBLEM: The first day in the field, the crew picked 400 quarts of strawberries. The second day they picked 320 quarts. What was the percentage of decrease in the number of quarts picked on the second day?

METHOD: (a) $400 - 320 = 80$

(b) $\dfrac{80}{400} = \dfrac{1}{5}$

(c) $\dfrac{1}{5} = 20\%$ ANS.

*Step 1*. Find the difference.
*Step 2*. Write a fraction of the amounts being compared.
*Step 3*. Express the fraction as a percent.

### Practice Exercise No. 73

Find the percent of increase or decrease in each pair—mentally.

1. 30 to 35
2. $1500 to $1800
3. 7 to 14
4. $2.00 to $3.50
5. 50 to 75
6. $8 to $6
7. 25 to 15
8. $12 to $3
9. $4.50 to $3.00
10. 200 to 100

### Practice Exercise No. 74

Solve the two-step percentage problems below.

1. A year ago Mr. Hadey's weekly salary was $450. All the employees received an increase. His salary is now $495. What percentage raise did he receive?

2. In the first month of publication, a best selling novel sold 50,000 copies. The next month the sale was 48,000 copies. What was the percentage decrease in the sales?

3. School bus fares two years ago were $3 a month. They are now $3.50 per month. What is the percent of fare increase?

4. If your household rent had been increased from $380 monthly to $456 monthly, what would be the percentage increase?

5. Mitch started working as a soda fountain clerk earning $3.80 an hour. He was given a $12\frac{1}{2}\%$ raise. What were his new hourly earnings?

6. Tom spends $50 a month on his car. He earns $8000 a year. What percent of his annual salary does he spend on the car?

7. Larry made $30 from his paper route, $15 from magazine subscriptions, $8 from mowing lawns, and $7 from washing cars. What percent of his earnings came from the paper route?

8. In a basketball foul shooting contest for accuracy, Bob dropped in 75 out of 150 tries. Jim was successful with 80 out of 175. Who had the better percentage and by how much?

9. Out of a junior chamber of commerce membership of 950, there was an attendance fo 725 at the annual convention. What percent of the membership did not attend?

10. After using his bike for six months, Russ sold it for 15% less than it cost. He sold it for $51. What did the bike cost originally?

# CHAPTER EIGHT

# HOW PERCENTS ARE USED IN DAILY BUSINESS

## DISCOUNT—COMMISSION—PROFIT AND LOSS

The arithmetic principles learned in handling percents play an important part in the daily business procedures concerned with the practices of Discounts, Commissions, Interest and Profit and Loss.

All of us come in contact with these business practices very early in our lives. You go to the department store and learn about discounts. Your family buys a home and someone talks about the real estate man's commission. At the bank they advertise the latest interest rates. The business proprietor and the manufacturer have to think about profit and loss at all times.

In order to understand these processes which are so common to everyday living, it is necessary to become familiar not only with the mechanics of the arithmetic, but also with the meanings of the special terms used in each field.

## DISCOUNT

To attract customers or build good will, a dealer reduces the price of an article or the amount of a bill. This is a **discount.**

A manufacturer sells goods to dealers, warehouses and jobbers for resale at reduced prices. The reductions in price offered to these merchants are termed **trade discounts.** The reductions are from the prices as advertised or listed in the manufacturer's catalogs. All such discounts are generally expressed as a *rate* or percent in relation to the original prices.

### Language of Discount

| | |
|---|---|
| **List price** <br> **Marked price** <br> **Former price** | These terms are used to denote the original price. |
| **Discount** or **Reduction** | This is the amount deducted from the original price. |
| **Net price** <br> **Sales price** | This is the lowered price. **To find net price,** *subtract the discount from the list price.* |

### Rate of Discount

**Rate of discount** is the percent represented by the amount of the discount in relation to the list price.

**Rule: To find the rate of discount,** *determine what fractional part the discount is of the original price and convert to a percent.*

**Rule: To find the amount of a discount,** *multiply the list price by the rate of discount.*

EXAMPLE 1: On sale, a fish tank was advertised at 20% off list price. The original tag read $5.50. What was the discount and the net price?

METHOD (a):

1. List Price           $5.50
   Rate of Discount    × .20
   Discount           $1.1000

2. List Price           $5.50
   Less Discount      − 1.10
                   $4.40 net price

*Step 1.* To find discount, multiply list price by *rate of discount*.

*Step 2.* To find net price, subtract discount from list price.

METHOD (b): An alternative method.

1. $100\% - 20\% = 80\%$

2. 80% of $5.50 = ?

$$\frac{4}{5} \times \$5.50 = \$4.40 \text{ net price.}$$

*Step 1.* Subtract rate of discount from 100%.

*Step 2.* Take this percentage of the list price.

This second method is often easier when Step 1 can be done mentally. This method is especially helpful in working out successive discounts.

EXAMPLE 2: A baseball catcher's mitt had a sign on it saying "Formerly $27, Reduced to $18." What is the discount and the rate of discount?

METHOD:

$$\$27 - \$18 = \$9 \text{ discount}$$

$$\frac{9}{27} = \frac{1}{3}$$

$$\frac{1}{3} = 33\frac{1}{3}\% \text{ rate of discount}$$

*Step 1.* Subtract sale price from list price.

*Step 2.* Find what fractional part the discount is of the list price and convert to a percent.

### Practice Exercise No. 75

Following are some offerings from the local daily newspaper. Fill in the missing quantities (to nearest whole percent).

| Article | Rate of Discount | Discount | List Price | Price |
|---|---|---|---|---|
| 1 Cashmere coat | | | $379.00 | $239.00 |
| 2 Sofa | | $156.00 | $749.00 | |
| 3 Shoes | | | $ 36.95 | $ 31.95 |
| 4 Swim suit | | | $ 19.95 | $ 12.95 |
| 5 Lamp | | | $ 16.95 | $ 9.95 |
| 6 Table | | | $ 49.98 | $ 37.98 |
| 7 Sweater | $\frac{1}{3}$ off | | $ 26.00 | |
| 8 Airplane fare | 5% | | $242.00 | |
| 9 Chair | | $ 27.50 | $ 97.50 | |
| 10 Rug | 25% | | $680.00 | |

### Practice Exercise No. 76

Solve the following problems pertaining to discounts.

**1.** A wholesaler states on its billing invoice that a 2% discount is allowed if the bill is paid within 10 days. The billing is for $77. How much is the discount and the final bill if paid in five days?

**2.** At a sporting goods store where Ed works, employees are given a 20% discount. What must he pay for a baseball bat marked $12.50?

**3.** An appliance store offered a discount of $33\frac{1}{3}\%$ on its floor models. What is the net price of a TV set listed at $240?

**4.** At the day-old bake shop yesterday's bread marked $1.25 was sold for $.75. What is the rate of price reduction?

**5.** The Discount Center offered a 20% reduction on an electric iron marked $38.00. What was the net price?

**6.** A store held a "Pre-Inventory Sale" on pre-teen dresses at $5, regularly $10.98 to $17.95. What was the rate of discount to the nearest percent on those marked $10.98 and $17.95?

**7.** The household furnishings department advertised a year-end clearance "regardless of cost; end tables regularly $45, now $19." What is the rate of discount to the nearest percent?

**8.** Ann bought a cultured pearl necklace at a 30% discount. It was marked $49.75 originally. A 10% tax was added to the selling price. What did she pay to the nearest penny?

**9.** At a storewide "10% off sale," Donald's father bought a fishing rod marked $29.95. What did he pay? Disregard fractions of a cent.

**10.** Alex was saving for a bicycle. On sale a $142.50 bike was offered at a 10% discount. Alex had $137.00 in his account. Could he buy the bicycle? How much over or under did Alex have in his account?

### To Find Original Price When Net Price Is Known

EXAMPLE: The ad read, "Camera on sale at $45, reduced 40%." What was the original price?

(This problem is the same as finding the whole when a percent is known.)

METHOD 1:

$$100\% - 40\% = 60\%$$

If $45 is 60%

$$\$45 \div \frac{60}{100} \text{ is } 100\%$$

$$\$45 \times \frac{5}{3} = \$75 \text{ Ans.}$$

*Step 1.* Subtract rate of discount from 100% to learn what percent of the whole the $45 represents.

*Step 2.* Find 60% of what = $45.

To check this answer, see whether a discount of 40% from $75 will give a net price of $45. Take $\frac{2}{5}$ of $75 and deduct. What do you get?

METHOD 2: Remember the 1% method for finding the whole when a percent is given.

$$100\% - 40\% = 60\%$$

If $45 is 60% then 1% would be ?

If 1% is ___ then 100% = ___ × 100 or $75.

#### Practice Exercise No. 77

Solve the problems below. They will test your ability to use discounts to find the original price.

**1.** Phil bought swim trunks at 15% off. He paid $4.00. What was the original list price?

**2.** Jerry bought a portable radio for $22 at a 20% discount. What was its list price?

**3.** Arthur bought a shirt at a discount of 18%. He paid $18 for it. What was the price of the shirt before it was reduced?

**4.** At the end of the summer Kenny bought a tennis racket for $10.50 which he was told represented a 25% discount. What was the original price?

**5.** Lenny bought a damaged tool kit at a discount of $33\frac{1}{3}$%. If he paid $22 for it, what was its original price?

### Figuring Multiple Discounts

In some business situations it is a practice to give more than one discount. For example, a dealer may offer a discount of 25% if an outlet buys 100 dozen articles of merchandise and an additional 10% if 200 dozen are purchased.

In some other cases, an extra discount is given above a regular discount if payment is made upon delivery or within a short period after delivery.

Discounts granted in this way are referred to as "chain discounts" or "successive discounts."

In figuring such chain or successive discounts, it should be noted that the second discount is figured on the net price after deducting the first discount. In the same way, a third discount in any chain is taken on the net after the two previous discounts are taken.

EXAMPLE 1: The A-One TV set listed at $200. The manufacturer allowed a mail-order firm trade discounts of 20% and 10% because they ordered two carloads. What was the net price per TV set?

METHOD (a):

20% of $200 = $\frac{1}{5}$ × $200 = $40 first discount.

$200 − $40 = $160 first discounted price.

10% of \$160 = $\frac{1}{10}$ of \$160 = \$16 second discount.

\$160 − \$16 = \$144 net price.

METHOD (b):
This is a variation of method (a).

10% of \$200 = $\frac{1}{10}$ of \$200 = \$20.

\$200 − \$20 = \$180.

20% of \$180 = $\frac{1}{5}$ of \$180 = \$36.

\$180 − \$36 = \$144 net price.

EXAMPLE 2: Take a 30% discount on \$200.

$$\frac{3}{10} \times \$200 = \$60$$

\$200 − \$60 = \$140 net price

Would you rather have successive discounts of 20% and 10% or a single discount of 30%?

You will observe from methods (a) and (b) that it does not matter whether the larger or the smaller discount is taken first. The net price is the same in both cases.

Would this hold true if you had three discounts of 20%, 10%, and 5% and varied the order? Try it.

Is a single discount of 35% the same as successive discounts of 20%, 10%, 5%? Which is a greater discount?

**Rules:** *Successive discounts cannot be added together.*

*The order in which successive discounts are taken does not affect the net price.*

**A shorter way to do successive discount** problems is to subtract the percentage of discount from 100% and multiply by the result.

EXAMPLE 3: A hi-fi set was listed at \$240 with discounts of 25% and 10%. What is the net price?

Method (b) is used here.

100% − 25% = 75%

75% of \$240 or $\frac{3}{4} \times \$240 = \$180$ first discounted price.

100% − 10% = 90%

90% of \$180 or $\frac{9}{10} \times \$180 = \$162$ net price.

**A still shorter way** to do successive discount problems is to do all the steps at the same time. Let's take Example 3 above.

There are 3 members to multiply: original price × first discount × second discount.

$$\overset{\overset{\overset{6}{\cancel{60}}}{\cancel{\$240}}}{} \times \frac{3}{\underset{1}{4}} \times \frac{9}{\underset{1}{\cancel{10}}} = \$162 \text{ net price}$$

or \$240 × .75 × .9 = \$162 net price

### Practice Exercise No. 78

Find the net price in the problems below.

| | List Price | Discounts | Net Price |
|---|---|---|---|
| 1. | \$300 | 20%, 10% | ? |
| 2. | \$400 | 15%, 15% | ? |
| 3. | \$200 | 30%, 2% | ? |
| 4. | \$500 | 10%, 5%, 2% | ? |
| 5. | \$250 | 40%, 20% | ? |

## COMMISSION OR BROKERAGE

Some people are paid a part of the money that results from a sale, a purchase or a collection instead of a fixed salary or hourly wage. Such payment is called a **commission.**

Payment on a commission basis is customary in many areas of the day-to-day business world.

Many salesmen are paid on a full commission or part commission, part salary basis as an incentive for them to work to their fullest abilities.

A farmer may employ an agent or a broker to sell his crops and pay him a commission for his services.

A man buying or selling stocks listed on the exchanges of New York, the Midwest or the Pacific Coast, uses a broker to do the purchasing or selling and is charged a commission.

A family selling a home usually places it with a real estate agent to whom it pays a commission or brokerage fee.

### Language of Commission and Brokerage

| | |
|---|---|
| **Sales Volume** **Selling Price** **Gross Proceeds** **Base** | These words are used to describe the money received by the representative for his employer. |
| **Commission** **Brokerage** | These terms apply to the *amount* of money the agent or broker receives. |

The **net proceeds** is the amount the employer finally receives. It is found by deducting the commission from the gross proceeds.

The **rate of commission** is the *percent* represented by the amount of commission in relation to the selling price, volume or gross proceeds. It is computed in the same way as the rate of discount.

Except for the variation of terms, the arithmetic of commission is exactly the same as the arithmetic you learned in finding percents and discounts.

**To find the rate of commission,** *determine what fractional part the commission is of the gross proceeds and convert to a percent.*

**To find the amount of commission or brokerage,** *multiply the principal amount (gross proceeds, etc.) by the rate of commission.*

EXAMPLE 1: Bill went out selling Home Chemistry Kits for the H-C-K Company on a 15% commission basis. He sold $275 worth of kits the first day. What was his commission? What was the net proceeds to the company?

METHOD (a):

| Sales | $275 |
|---|---|
| Rate of Commission | × .15 |
| Commission | $41.25 or |

$$\$275 \times \frac{15}{100} =$$

$$\$275 \times \frac{3}{20} = \$41.25$$

*Step 1.* To find commission, multiply gross proceeds by rate of commission.

| Sales | $275 |
|---|---|
| Less Commission | − 41.25 |
| Net Proceeds | $233.75 |

*Step 2.* To find net proceeds subtract commission from sales volume (gross proceeds).

METHOD (b):

$$100\% - 15\% = 85\%$$

$$\$275 \times \frac{85}{100} = \$233.75 \text{ Net proceeds}$$

$$\$275 - \$233.75 = \$41.25 \text{ Commission}$$

*Step 1.* Subtract the rate of commission from 100%.

*Step 2.* Take this percentage of the gross sales.

*Step 3.* Gross sales less net proceeds equals amount of commission.

In doing problems about commission, proceed as the salesman or agent does. They like to estimate their commission earnings. In this way they know how much to expect and it acts as a check on the computations that may be involved.

Estimate the answer before solving.

EXAMPLE 2: A real estate agent earned a commission of 5% for selling a lot at a sales price of $22,500. What was his commission?

Estimate—Consider the amount to be $22,000. Think—10% of $22,000 equals $2200. 5% is half of 10%. $\frac{1}{2}$ of $2200 is $1100. Estimated answer.

METHOD:

$ 22,500
$\times$ .05

$1125.00    Actual answer is reasonable.

### Practice Exercise No. 79

Do the following problems relating to commission.

**1.** Peter's father is a furniture salesman who receives a 3% commission on his total sales volume. His sales during the week showed $400 for Monday, $575 for Tuesday, $550 for Wednesday, $800 for Thursday, $250 for Friday and $1200 on Saturday. What was his commission for the week?

**2.** A real estate agent sold a house for $125,400. He was not a licensed broker and had to split the regular 5% brokerage fee on a fifty-fifty basis with the office out of which he worked. What was his commission?

**3.** Working on a straight 2% commission basis Mr. Molitini sold 45 sets of plated silverware at $95.50 a set. What was his commission?

**4.** Leon's mother works in the cosmetic section of the local department store where she receives a salary of $160 a week plus a commission of $1\frac{1}{2}$% on sales. Her sales for the week were $1830. How much did she earn for the week?

**5.** Allan's father works as a route salesman for a baking company. He receives a base salary of $150 per week, a commission rate of 12% on bread and rolls and 15% on cakes or sweet goods. His sales for the week were $825 in bread and rolls and $750 in sweet goods. How much did he earn?

**6.** A real estate broker sold a house for $215,500. His commission arrangement was 5% for the first $100,000 and 3% for any amount over it. What was his commission?

**7.** Floyd sold a car for Mr. Janus for $1800. He received $72 as his commission fee. What was his rate of commission?

**8.** Clara sold $70 worth of magazines in one week and received $14 as her commission. What was the rate of commission?

## MATHEMATICS OF BUYING AND SELLING OR PROFIT AND LOSS

Unlike the commission-type payment, we have the earnings of the man who owns a business, he may be the retail storekeeper, the farmer, the manufacturer, the wholesaler, the supplier of services. His earnings have to be figured on the cost of his products along with the cost of doing business or "overhead expense" as it is called. After paying for goods, commissions, rents, utilities and other expenses he will have what is generally called a profit or loss.

The way in which profit and loss is computed is an important part of the arithmetic of everyday business. In some cases it is quite simple as when an individual buys one item for immediate resale. In other cases, such as large manufacturing, farming, lumbering or any type of big business, the figuring of profit and loss statements is a highly specialized kind of mathematics which needs to be performed by a trained accountant.

For our purpose, we will take up the more simple type of profit and loss situations that everyone of us may come in contact with from time to time.

EXAMPLE 1: Mr. Rick sold 100 hammers at $1.00 each. He paid 75 cents per hammer.

What terms do we find in this simple problem of profit and loss?

The cost or **first cost** is the payment for goods made by the businessman before adding on expenses. Mr. Rick's first cost for the hammers was $75.

The **selling price** is the amount received by the businessman for the goods he sells. Mr. Rick's selling price for the hammers was $100.

The **gross profit** is the difference between the selling price and the first cost.

In this case, the gross profit is:

$$\$100 - \$75 = \$25$$

Now we ask what is Mr. Rick's margin of profit?

When expressed as a percent, this margin of profit is called the **markup.**

*Markup* can be taken as a percent of the cost or of the selling price. For example:

$$\frac{\$.25}{\$.75} = 33\tfrac{1}{3}\% \text{ markup if taken on the cost per hammer.}$$

$$\frac{\$.25}{\$1.00} = 25\% \text{ markup if taken on the selling price per hammer.}$$

It should be noted that according to modern accounting and department store management practices the idea of basing the markup on the selling price is favored. The reason for this is the fact that commissions and other selling expenses are figured as percentages of selling price. Thus it simplifies such accounting to base profit and loss on the selling price also. However, many people in business continue to figure their markup as a percentage of *cost*. Therefore we shall practice both.

The previous example is oversimplified because we made no mention of overhead expense. Actually we concerned ourselves only with the cost and therefore could only compute the *gross profit*.

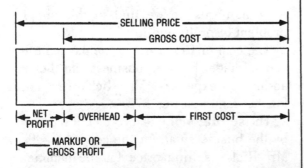

A profitable business

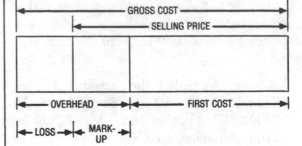

An unprofitable business. In an unprofitable situation, notice how the *loss* is due to the overhead, even though the goods are sold above the purchase price.

### Practice Exercise No. 80A

Jack sold Christmas cards one year. He placed ads to attract customers. He purchased boxes of assorted cards at 44¢ per box and sold them for 66¢ per box. In one month he sold $204.60 worth of cards and spent $13.64 on ads, postage and telephone calls. Answer the questions, doing all the necessary arithmetic.

**(a)** How many boxes of cards were sold?
$$\$204.60 \div ? = ?$$

**(b)** What was the first cost of all the boxes sold?
$$44¢ \times ? = ?$$

**(c)** How much *net* profit was made?
$$\$204.60 - \$13.64 - ? = ?$$

**(d)** What was the amount of the markup per box?
$$66¢ - ? = ? \text{ markup}$$

**(e)** The markup was what percent of the cost per box?
$$\frac{?}{44¢} = ? = ? \text{ expressed as a percent}$$

**(f)** The markup was what percent of the selling price per box?
$$\frac{?}{66¢} = ? = ? \text{ expressed as a percent}$$

**(g)** The net profit was what percent of the gross cost?
$$\frac{\$204.60 - \$13.64 - ?}{44¢ \times \text{No. of boxes} + \$13.64} = ? \text{ expressed as a percent}$$
$$\phantom{\frac{}{\text{sold}}}\text{sold}$$

**(h)** The net profit was what percent of the selling price?
$$\frac{\$204.60 - \$13.64 - ?}{\$204.60} = ? \text{ expressed as a percent}$$

**Practice Exercise No. 80B**

Fill in the blanks with the correct word or words.
EXAMPLE: First cost + markup = selling price.

1. First cost + overhead + profit = _____

2. Markup + _____ = selling price

3. Overhead + first cost = _____

4. Markup *less* profit = _____

5. Selling price *less* profit = _____

6. Gross cost *less* first cost = _____

7. First cost + overhead *less* loss = _____

8. Overhead *less* markup = _____

9. Selling price *less* _____ = profit

10. Gross cost = first cost + _____

In the problems which follow, we will not consider overhead but will concentrate on giving you practice in applying percent to problems of buying and selling.

EXAMPLE 2: A shirt that cost $18 was sold by the haberdasher for $22.50. What was the gross profit? What was the rate of profit (markup) based on *cost*? What was the markup based on selling price?

METHOD:

1. $\begin{array}{r} \$22.50 \\ -\ 18.00 \\ \hline \$\ 4.50 \end{array}$

2. $\dfrac{\text{Gr. Profit}}{\text{Cost}}\ \dfrac{\$\ 4.50}{\$18.00} = \dfrac{1}{4} = 25\%$ based on cost

3. $\dfrac{\text{Gr. Profit}}{\text{Selling Price}}\ \dfrac{\$\ 4.50}{\$22.50} = \dfrac{1}{5} = 20\%$ based on selling price

*Step 1.* Gross profit = selling price less cost.

*Step 2.* Percent of profit equals amount of profit divided by cost.

or

*Step 3.* Percent of profit equals amount of profit divided by selling price.

EXAMPLE 3: An appliance dealer obtained a radio for $15 and sold it at a price 40% above the cost. What was the gross profit? What was the selling price?

METHOD:

1. 40% of $15 =

   $\dfrac{2}{5} \times \$15 = \$6.00$ gross profit

2. $\begin{array}{r} \$15 \\ +\ 6 \\ \hline \$21 \end{array}$ selling price

*Step 1.* To find gross profit on cost, multiply cost by markup (percent profit).
*Step 2.* To find selling price, add the gross profit to the cost.

EXAMPLE 4: A hobby show owner sold a puppy to a young customer he liked at 20% less than its cost of $6.00. What was his loss? What was the selling price?

1. 20% of $6 =

   $\dfrac{1}{5}$ of $6 = $1.20 loss

2. $\begin{array}{r} \$6.00 \\ -\ 1.20 \\ \hline \$4.80 \end{array}$ selling price

*Step 1.* To find profit or loss, multiply cost by percent "markup" or "markdown."
*Step 2.* To find selling price, subtract the loss from the cost.

**Practice Exercise No. 81**

Using the information given, fill in the empty boxes.

|  | Cost | Gross Profit | % Profit on Cost | % Profit on Selling Price | Selling Price |
|---|---|---|---|---|---|
| 1. | $75 | $25 |  |  |  |
| 2. |  |  |  | 20% | $500 |
| 3. | $200 |  |  |  | $300 |
| 4. | $160 |  | 25% |  |  |
| 5. | $1450 |  |  |  | $2000 |

## Practice Exercise No. 82

In Problem 1 overhead is included. Problem 1A is done as a sample. Before proceeding with Problems 1B-1E read the explanation below. Problems 2-11 will test your knowledge of profit and loss.

**1.**

| | First Cost | Selling Price | Overhead | | Markup | | Profit | | |
|---|---|---|---|---|---|---|---|---|---|
| | | | % of Cost | Dollars | % of Cost | Dollars | In dollars | In % of Sales | In % of Cost |
| **A.** | $300 | **$500** | 50% | **$150** | $66\frac{2}{3}$ | $200 | **$50** | 10% | $16\frac{2}{3}\%$ |
| **B.** | $150 | ? | 40% | ? | ? | $120 | ? | ? | ? |
| **C.** | $250 | ? | ? | $100 | ? | $225 | ? | ? | ? |
| **D.** | ? | $290 | ? | ? | ? | $140 | ? | 20% | ? |
| **E.** | $750 | ? | ? | $150 | ? | $250 | ? | ? | ? |

EXPLANATION OF PROBLEM 1A

We know that First Cost + Markup gives us the Selling Price. Thus $300 + $200 = $500 (Selling Price).

Overhead, in dollars + Profit, in dollars = Markup. We can find the overhead, because we have % of overhead as 50% of cost or $\frac{50}{100} \times \$300 = \$150$. Now we can find the Profit in that Overhead + Profit = Markup. Thus $150 + ? = $200. Profit is $50. Now we can proceed to find the remaining answers as follows:

$$\text{Markup as a \% of cost } \frac{\$200}{\$300} = \frac{2}{3} = 66\frac{2}{3}\%$$

$$\text{Profit as a \% of sales } \frac{\$50}{\$500} = \frac{1}{10} = 10\%$$

$$\text{Profit as a \% of cost } \frac{\$50}{\$300} = \frac{1}{6} = 16\frac{2}{3}\%$$

**2.** Hector bought a bike for $100. It cost him $35.00 to fix it. He resold it at a net profit of $55. What was the selling price? What was his rate of profit on the selling price?

**3.** Mr. Baker studied his expenses in the running of his camera exchange shop and found his overhead to be about 30% of the cost of the products he sells. To begin with he decided to add 30% to the cost of every item. He further decided that he could add another 20% to the cost of the materials as his profit and still be competitive. What price would he then have to charge for a small camera that costs him $8.00? How much net profit will he make on it?

**4.** What is the selling price of a banjo that cost the dealer $250 if his overhead runs 45% of his merchandise costs and if he works on a profit of 20% of cost?

**5.** A grocer bought a carton of 24 cans of fruit juice for $6.00. His overhead is figured at $33\frac{1}{3}\%$ of his cost. He desires a profit of 10% of his cost. What must be his selling price *per can* of juice? (To nearest cent.)

**6.** A dealer bought a lead-item typewriter for $235. He added $90 for overhead and sold it for $475. What was his profit and what was his rate of profit?

**7.** A table that sells for $175 cost the dealer $100. He figures his overhead to be $30 and his profit $45. What percent of his cost is his overhead? What is his percent of profit based on the selling price?

**8.** Arvid bought fish hooks at 48¢ a dozen and sold them to his friends for 5¢ each. Based on the selling price what was his percent of markup?

**9.** Mr. Allen wants a line of suits to retail for $140. His overhead plus profit is to be 40% based on selling price. What price should he pay per suit?

**10.** Mary Dee made copper trinkets and sold 25 of them at a price of 88¢ each. Her materials cost her $8.00. What was her percent of markup based on the selling price?

**11.** Mr. Kahn had a good day in his teenage shop. He grossed $2000. His accountants figured his overhead at 25% of his dollar sales volume. He worked at a $37\frac{1}{2}$% markup based on selling price. What was his overhead? What was his profit for the day?

# CHAPTER NINE

# HOW MONEY IS USED TO EARN MONEY

## FIGURING INTEREST AND BORROWING

Harold Blatz borrowed $400 from his neighbor. He agreed to pay his neighbor 5% for the use of the money. He promised to repay the loan and the 5% at the end of a year. How much would he have to give his neighbor at that time?

This is an elementary problem in computing simple *interest*. Problems in interest require you to use most of the procedures you learned in studying percentage but include the additional factor of *time*. Interest is a way in which money earns money.

## DEFINITIONS

Like percentage, discount, and commission, the study of interest has its special terms.

The **interest** $(I)$ is the charge for the use of money.

The **principal** $(P)$ is the money borrowed on which interest is paid.

The **rate** of interest $(R)$ is the percentage charged on the basis of one year's use of the money.

The **time** $(T)$ is the number of years, months and days during which the money is used.

The **amount** $(A)$ is the sum of the principal and the interest.

In the problem given:

Principal = $400
Rate = 5%          $I = ?$
Time = 1 year

The interest is found by carrying out the same processes we use for any other percent problem.

To find the interest, we take 5% of the principal.

$$I = \$400 \times .05 = \$20 \text{ or}$$

$$\frac{\$400}{1} \times \frac{5}{100} = \$20$$

The amount $(A)$ to be repaid = $400 + $20 = $420

In this case, the time $(T)$ was one year and was not used as an additional multiplier. If the loan had been taken for two years, we would have multiplied the interest ($20) by 2.

With reference to time, note that 30 days are considered a month and 360 days are considered a year in figuring interest charges.

## FINDING INTEREST BY FORMULA

**Rule: To find the interest for any given period of time,** *multiply the principal by the rate by the time.*

**Formula:** $I = P \times R \times T$.

EXAMPLE 1: Find the simple interest and amount to be paid on $800 at 6% for 3 years.

$$I = \$800 \, (P) \times .06 \, (R) \times 3 \, (T)$$

$$I = \frac{\overset{8}{\cancel{\$800}}}{1} \times \frac{6}{\underset{1}{\cancel{100}}} \times 3 = \$144 \text{ Ans.}$$

or

$$I = \begin{array}{r} \$\ 800 \\ \times\ .06 \\ \hline \$48.00 \end{array} \qquad \begin{array}{r} \$\ 48 \\ \times\ 3 \\ \hline \$144 \end{array}$$

Amount =
$800P + $144I = $944 Ans.

EXAMPLE 2: Find the amount that is to be repaid if $500 is borrowed for 3 years 3 months and 15 days at 4% (3 months is considered 90 days, $\frac{90}{360}$ of a year or $\frac{3}{12} = \frac{1}{4}$ of a year). We therefore find the interest for $\frac{1}{4}$ of a year.

METHOD: $I = P \times R \times T$

$500 \times .04 \times 3 \quad = $60.00

$500 \times .04 \times \frac{1}{4} \quad = $ 5.00

$500 \times .04 \times \frac{15}{360} = \quad .83\frac{1}{3}$

$\overline{\quad\$65.83\quad}$

Amount = $P + I$.
$500 + $65.83 = $565.83 amount.

### Practice Exercise No. 83

Given the principal, rate and time, find the interest and amount in the examples which follow.

| | Principal | Rate | Time | Interest | Amount |
|---|---|---|---|---|---|
| 1. | $575 | 2% | 1yr. | ? | ? |
| 2. | $200 | 5% | 1yr. | ? | ? |
| 3. | $350 | 3% | 1yr. | ? | ? |
| 4. | $550 | 6% | 1yr. | ? | ? |
| 5. | $400 | 4% | 1yr. | ? | ? |
| 6. | $850 | 6% | 1yr. | ? | ? |
| 7. | $1200 | 3% | 1yr. | ? | ? |
| 8. | $900 | 5% | 1yr. | ? | ? |
| 9. | $390 | 4% | 1yr. | ? | ? |
| 10. | $1500 | 2% | 1yr. | ? | ? |
| 11. | $600 | $4\frac{1}{2}$% | 1yr. | ? | ? |
| 12. | $450 | $2\frac{1}{2}$% | 1yr. | ? | ? |
| 13. | $900 | $5\frac{1}{2}$% | 1yr. | ? | ? |
| 14. | $370 | $3\frac{1}{2}$% | 1yr. | ? | ? |
| 15. | $550 | $4\frac{1}{2}$% | 1yr. | ? | ? |
| 16. | $1200 | $1\frac{1}{2}$% | 1yr. | ? | ? |

### Practice Exercise No. 83 *(Continued)*

| | | | | | |
|---|---|---|---|---|---|
| 17. | $800 | $4\frac{1}{4}$% | 1yr. | ? | ? |
| 18. | $300 | $5\frac{1}{2}$% | 1yr. | ? | ? |
| 19. | $1000 | $2\frac{1}{4}$% | 1yr. | ? | ? |
| 20. | $500 | $1\frac{1}{2}$% | 1yr. | ? | ? |
| 21. | $500 | 6% | 8 mos. | ? | ? |
| 22. | $250 | 5% | 6 mos. | ? | ? |
| 23. | $400 | 5% | 2 yrs. | ? | ? |
| 24. | $600 | 4% | 3 mos. | ? | ? |
| 25. | $400 | 2% | $1\frac{1}{2}$ yrs. | ? | ? |
| 26. | $300 | 4% | 2 yrs. 2 mos. | ? | ? |
| 27. | $175 | 6% | 4 mos. | ? | ? |
| 28. | $1500 | 3% | 2 mos. | ? | ? |
| 29. | $200 | 2% | 9 mos. | ? | ? |
| 30. | $800 | 4% | 15 mos. | ? | ? |

## INDIRECT CASES OF INTEREST

**Rule: To find the rate when the principal, interest and time are given,** *divide the total interest by the time to get the amount of interest for one year;* then *divide this quotient by the principal.*

EXAMPLE 1: What must be the rate of interest on $800 to produce $30 in $1\frac{1}{2}$ years?

METHOD:

$30 \div \frac{3}{2} = $30 \times \frac{2}{3} = $20$ interest for 1 yr.

$20 \div $800 = \frac{2}{80} = \frac{1}{40} = 2\frac{1}{2}$% rate of interest

To check: Take $2\frac{1}{2}$% of $800 for $1\frac{1}{2}$ yrs.

**Rule: To find the time, when the principal, interest and rate percent are given,** *multiply the principal by the rate to obtain the interest for one year;* then *divide the total interest by the interest for one year.*

EXAMPLE 2: How long will it take for $1200 to yield $60 in interest at a rate of 4%?

METHOD:

$1200 × .04 = $48.00 interest for 1 yr.

$$\frac{\$60}{\$48} = \frac{5}{4} = 1\frac{1}{4} \text{ years Ans.}$$

### Practice Exercise No. 84

Solve the interest problems below.

**1.** What must be the rate of interest on $600 to produce $18 in $1\frac{1}{2}$ yrs.?

**2.** How long will it take for $500 to yield $60 at 6% interest?

**3.** What rate of interest should be charged on $900 to earn $22.50 in 1 year?

**4.** For how long must I lend $700 at 5% interest to earn $52.50?

**5.** What rate of interest is required to produce $120 in two years at simple interest when the principal is $1500?

**6.** To earn $24 at 4% simple interest, how long must $900 be invested?

**7.** What must be the interest rate on $1600 to earn $40 in 6 months?

**8.** How long will it take for $1100 at 3% interest to yield $82.50?

**9.** With $1400 to invest for 8 months, what rate is needed to earn $56.00?

**10.** With $1300 to invest at 2% simple interest, how long will it take to earn $78?

### THE 60 DAY-6% METHOD OF FIGURING INTEREST

As noted before, in figuring interest, the banks and many businesses use a 360-day year divided into 12 months of 30 days each.

On this basis, it is convenient to use what is known as the 60 day-6% method of figuring interest.

Thus interest on $1.00 for 1 year at 6% is six cents (.06). For 60 days it is $\frac{1}{6}$ or one cent (.01). From this fact that interest on $1.00 for 60 days at 6% is one cent, we observe that *the interest on any amount of money at 6% for 60 days is 1% of the principal.*

Thus:

Interest on $140 at 6% for 60 days = $1.40

Interest on $355 at 6% for 60 days = $3.55

Observing what takes place we may derive a useful rule for this so-called 6% method:

**To find interest for 60 days at 6%,** *move the decimal point in the principal two places to the left.*

### Practice Exercise No. 85

Using the 6% method, find the interest on the following for 60 days at 6%.

| | |
|---|---|
| **1.** $340 | **6.** $111.30 |
| **2.** $865 | **7.** $91.80 |
| **3.** $1450 | **8.** $642.50 |
| **4.** $30 | **9.** $1.20 |
| **5.** $921 | **10.** $4259.30 |

### Applying the 60 Day—6% Method to Other Terms

It is often convenient to convert varied periods of time and interests to fractions and multiples of the 60 day and 6% figure.

EXAMPLE 1: Find the simple interest on $630 at 6% for 96 days.

METHOD:

$ 6.30 = 60 days interest at 6% on $630
$ 3.15 = 30 days interest ($\frac{1}{2}$ of 60 days)
$  .63 =  6 days interest ($\frac{1}{10}$ of 60 days)
$10.08 = 96 days interest at 6% on $630

EXAMPLE 2: What is the interest on $320 for 30 days at 6%?

METHOD: $3.20 = 60 days interest at 6%.

$\frac{1}{2}$ of $3.20 = $1.60 = 6% for 30 days Ans.

EXAMPLE 3: What is the interest on $420 for 60 days at 4%?

METHOD: $4.20 = interest for 60 days at 6%.

Since 4% is $\frac{2}{3}$ of 6%, $\frac{2}{3}$ of $4.20 =
$2.80 ANS.

EXAMPLE 4: What is the interest on $400 at $1\frac{1}{2}$% for 4 months?

METHOD:
$4.00 = interest at 6% for 2 months
$8.00 = interest at 6% for 4 months

Since $1\frac{1}{2}$% is $\frac{1}{4}$ of 6%, $\frac{1}{4}$ of $8.00 =
$2.00 ANS.

### Practice Exercise No. 86

Using the 60 day—6% method try to find the answers mentally.

1. $500 for 2 months at 3%
2. $3000 for 3 months at 2%
3. $600 for 6 months at 3%
4. $1200 for 2 months at 1%
5. $750 for 4 months at 4%
6. $840 for 30 days at 3%
7. $5000 for 1 month at 6%
8. $1100 for 1 year at 3%

Use pencil and paper for the following:

9. $540 for 90 days at 5%
10. $210 for 120 days at $1\frac{1}{2}$%
11. $960 for 36 days at 3%
12. $1080 for 110 days at $3\frac{1}{2}$%
13. $1600 for 30 days at 2%
14. $560 for 150 days at 4%
15. $720 for 70 days at 7%
16. $628 for 90 days at 2%
17. $960 for 36 days at $4\frac{1}{2}$%
18. $840 for 72 days at $3\frac{1}{2}$%
19. $1840 for 63 days at 4%
20. $1960 for 12 days at $1\frac{1}{2}$%

## COMPOUND INTEREST

In the preceding problems the interest was added to the principal *at the end of the paying period*. This is termed **simple interest.**

There is another way in which interest is paid. This method is used by banks and is called **compound interest** to identify it.

Let us say that you deposit $150 in the Mutual Credit Association. This bank advertises that it pays 4% compounded quarterly. The words "compounded quarterly" mean that the interest is calculated every three months (or each quarter of a year) and added to the principal. Each addition to the principal in turn earns interest at following quarterly periods.

Here are the calculations that would show how the bank arrives at the entries you would find in your bank book if you left $150 in the bank for one full year.

ENTRIES:
$150 Principal for 1st period—1st three months
× .01 Interest rate for three months (4% per year)
───
$1.50 Interest for 1st period

$150 + $1.50 = 151.50 *New Principal*
$151 Principal for 2nd period (only *dollars* are considered by the bank)
× .01 Interest *rate* for three months (4% per year)
───
$1.51 Interest for 2nd period

$151.50 + $1.51 = $153.01 *New Principal*

$153 Principal for 3rd period
× .01 Interest *rate* for three months
$1.53 Interest for 3rd period

$153.01 + $1.53 = $154.54 *New Principal*

$154 Principal for 4th period
× .01 Interest rate for three months
$1.54 Interest for 4th period

$154.54 + $1.54 = $156.08—principal at the end of one year for $150 at 4% compounded quarterly.

Compare this with the result if the $150 had earned only simple interest at 4% for one year.

$150 × .04 = $6.00 @ simple interest versus $6.08 @ compound interest (quarterly).

With simple interest at 4% your money will be doubled in 25 years.

With compound interest at 4%, compounded annually, your money will double itself in approximately $17\frac{1}{2}$ years. If compounded semiannually or quarterly, it will take slightly less time to double itself.

### Practice Exercise No. 87

Find the total amount you would receive in each of the following examples. In all cases the interest is to be computed as compound interest.

1. On $100 for 2 years at 3% annually.

2. On $250 for 2 years at 4% semiannually.

3. On $500 for $1\frac{1}{2}$ years at 3% semiannually.

4. On $800 for 3 years at 2% annually.

5. On $650 for 4 years at 4% semiannually.

### Problems in Compound Interest

6. If Mickey put $500 in a savings bank at 6% interest payable semiannually, how much could he withdraw at the end of two years and still have $500 in the bank?

7. A neighborhood savings & loan association pays interest at 7% per annum. If you deposited $200 and left it there for three years, how much would you have at the end of that period?

8. How much would $1000 earn for you at the end of three years in a bank at $3\frac{1}{2}$% compounded semiannually?

9. At $5\frac{1}{2}$% compounded semiannually, how much will George get in interest at the end of five years on the $1500 he received in gifts for his 13th birthday?

10. If your parents set aside a trust fund of $2000 in a first mortgage for you that earned 6% and it was compounded semiannually, how much would you have at the end of 10 years?

## HOW TO USE A COMPOUND INTEREST TABLE

There is no doubt that the computations for the preceding exercises are lengthy and time consuming, even if they are not difficult. As you might expect, it would not be practical for banks, lending companies and others who have to figure such interest frequently to do the arithmetic each time. Actually, they have made the computations with the aid of computers and arranged the facts in convenient tables called Compound Interest Tables. A sample table follows.

### Compound Interest Table

One dollar invested at the % given is compounded quarterly.

| TIME IN YEARS | 4 | 4.5 | 5 | 5.5 | 6 | 6.5 |
|---|---|---|---|---|---|---|
| 1 | 1.040604 | 1.045765 | 1.050945 | 1.056145 | 1.061364 | 1.066602 |
| 2 | 1.082857 | 1.093625 | 1.104486 | 1.115442 | 1.126493 | 1.137639 |
| 3 | 1.126825 | 1.143674 | 1.160755 | 1.178068 | 1.195618 | 1.213408 |
| 4 | 1.172579 | 1.196015 | 1.219890 | 1.244211 | 1.268986 | 1.294222 |
| 5 | 1.220190 | 1.250751 | 1.282037 | 1.314067 | 1.346855 | 1.380420 |
| 6 | 1.269735 | 1.307991 | 1.347351 | 1.387845 | 1.429503 | 1.472358 |
| 7 | 1.321291 | 1.367852 | 1.415992 | 1.465765 | 1.517222 | 1.570419 |
| 8 | 1.374941 | 1.430451 | 1.488131 | 1.548060 | 1.610324 | 1.675012 |
| 9 | 1.430769 | 1.495916 | 1.563944 | 1.634975 | 1.709140 | 1.786570 |
| 10 | 1.488864 | 1.564377 | 1.643619 | 1.726771 | 1.814018 | 1.905559 |
| 15 | 1.186697 | 1.956645 | 2.107181 | 2.269092 | 2.443220 | 2.630471 |
| 20 | 2.216715 | 2.447275 | 2.701485 | 2.981737 | 3.290663 | 3.631154 |
| 25 | 2.704814 | 3.060930 | 3.463404 | 3.918201 | 4.432046 | 5.012517 |

To read the table, find the column that shows the rate of interest that is paid. Look down that column until you come to the row that reflects the number of years indicated in the particular problem. This number shows how $1 grows in principal and interest, when compounded quarterly, in that time.

EXAMPLE: What amount will we find for $1 deposited for 10 years at $4\frac{1}{2}$% compounded quarterly?

PROCEDURE: Look down the $4\frac{1}{2}$% column until you come to the 10-year row. The number is 1.56437. This means $1 will grow to $1.56 in 10 years. If the original amount had been $100, you would multiply by 100 to get $156.44.

How would you find the amount for $1 at the end of 20 years at the $4\frac{1}{2}$% rate? Continue down the $4\frac{1}{2}$% column to the 20-year row. What number do you see there? Is it 2.44727?

EXAMPLE: Using the table, how much would $2000 grow to in 5 years with interest compounded quarterly at 6%?

METHOD: Follow down the 6% column and opposite the 5-year row in the table, and we find that $1 compounded at 6% quarterly for 5 years is $1.34685.

$$\$2000 \times 1.34685 = \$2693.70$$

**Practice Exercise No. 88**

Use the Compound Interest Table to solve the problems which follow.

**1.** John's father placed $200 in a savings account when John was born. It has been earning 4% interest compounded quarterly for 15 years. How much does he now have in the bank?

**2.** Ten years ago Henry put $100 in a bank that compounded interest quarterly. If the bank book shows $190.56, what rate of interest does this bank pay?

**3.** Find the amount to which $1000 will grow if deposited at 5% compounded quarterly in
        (a) 2 years
        (b) 5 years
        (c) 15 years.

**4.** Mr. Jacobson deposited $400 in a bank that paid 6% interest compounded quarterly. He withdrew half the balance at the end of 4 years. How much money was left in the account?

**5.** Malvern won $2500 in a TV name contest. He put it in a bank that pays interest at the rate of 5.5% each 90 days. How much money will he have to use for tuition toward his college education at the end of 10 years?

## BORROWING MONEY

When money is borrowed from a bank or a lending institution, there are certain general practices observed, which you should know.

In borrowing money from an institution, the borrower usually has to sign a promissory note, a chattel mortgage or other binding legal form. A **promissory note** is a promise to repay a loan on a certain date.

The interest charges on a promissory note are *deducted in advance* from the principal or face value of the note. This is called **discounting** the note. The amount deducted is called the **discount.**

The **principal** or face value of the note is the amount borrowed.

The remainder left when the *discount* has been *deducted* from the *face* of the note is termed the **net proceeds** and is the amount the borrower receives.

At **maturity,** which is the date the loan is repayable, the borrower pays back the full face value of the loan. Let's study an example of a discounted loan.

EXAMPLE: Mr. Rieker borrows $300 from his local bank on a promissory note for 90 days at 6%. What are the amounts of the bank discount and the net proceeds?

METHOD:

$$\$300 \times \frac{6}{100} \times \frac{90}{360} = \text{Interest}$$

$$\frac{\overset{3}{\cancel{\$300}}}{1} \times \frac{6}{\cancel{100}} \times \frac{\overset{1}{\cancel{90}}}{\underset{4}{\cancel{360}}} = \frac{18}{4} = \$4.50$$
$$\text{discount}$$

or by 6%-60 day method
$$\$3.00 = 6\% \text{ for } 60 \text{ days}$$
$$\underline{+ \ 1.50} = 6\% \text{ for } 30 \text{ days}$$
$$\$4.50 = 6\% \text{ for } 90 \text{ days}$$
$$\$300 - \$4.50 = \$295.50 \text{ Net Proceeds}$$

In discounting loans, the *true rate* of interest is greater than the interest rate indicated. In the previous example, the interest paid is $4.50 for $295.50 for 90 days *not* for $300 for 90 days. If for example, one borrows $10,000 at the 6% rate for a full year, the *discount* is $600. This makes a sizable difference.

Compute the actual interest rate on a payment of $600 for $9400 for one year.

$600 is __% of $9400? (To the nearest tenth percent.)

### Practice Exercise No. 89

Find the discount and net proceeds on these loans.

| | Principal | Rate | Time Discount | Discount in Cash | Net Proceeds |
|---|---|---|---|---|---|
| 1. | $540 | 6% | 30 days | _____ | _____ |
| 2. | $350 | 6% | 60 days | _____ | _____ |
| 3. | $220 | 5% | 90 days | _____ | _____ |
| 4. | $200 | 4% | 45 days | _____ | _____ |
| 5. | $150 | 9% | 120 days | _____ | _____ |

## SMALL LOANS AND INSTALLMENT BUYING

Often a person needs small sums of money, but does not have tangible security to guarantee the loan. He may obtain a personal loan from a finance company. Such loans are repayable on a monthly basis in the form of installments rather than in a lump sum on a date of maturity.

This is the way most automobile and appliance installment buying is financed. The amount of the loan the first month is the amount of the unpaid balance plus the finance charges. Here is an example of such a loan.

EXAMPLE 1: Mr. Roberts borrows $100 from a bank to be repaid in 12 installments of $9.75. What rate of interest would he be paying on such a loan?

METHOD: In 12 months Mr. Roberts pays back

$$12 \times \$9.75 = \$117$$

Does this mean his rate of interest is $17/$100 or 17%? Actually it is much greater. You see, he did not borrow $100 for 12 months. He only borrowed $100 for the 1st month. Since he began to repay the loan the second month, he only had $90.25 of the bank's money. ($100 − $9.75) The third month he only had $80.50 of the bank's money. ($90.25 − $9.75) The fourth month $70.75 and so on until the total amount is paid.

To calculate the rate of interest on such a loan, we need to find the average amount that Mr. Roberts owed the bank over the 12 months. To find the average loan we take the amounts owed each month, add them up, then divide by the number of months.

Look at the table below, which is made by subtracting the amount of $9.75 continuously:

| Month | Amount of Loan | Payment | Amount Owed Bank |
|---|---|---|---|
| 0 | $100.00 | | $117.00 |
| 1 | $ 90.25 | $9.75 | $107.25 |
| 2 | $ 80.50 | $9.75 | $ 97.50 |
| 3 | $ 70.75 | $9.75 | $ 87.75 |
| 4 | $ 61.00 | $9.75 | $ 78.00 |
| 5 | $ 51.25 | $9.75 | $ 68.25 |
| 6 | $ 41.50 | $9.75 | $ 58.50 |
| 7 | $ 31.75 | $9.75 | $ 48.75 |
| 8 | $ 22.00 | $9.75 | $ 39.00 |
| 9 | $ 12.25 | $9.75 | $ 29.25 |
| 10 | $ 2.50 | $9.75 | $ 19.50 |
| 11 | $ 0 | $9.75 | $ 9.75 |
| 12 | $ 0 | $9.75 | $ 0 |

If we add the amounts in the *amount of loan* column we see the sum is $563.75.

$$\$563.75 \div 12 = \$46.81$$

So the average loan is $46.81. This means that Mr. Roberts only had the equivalent of the use of $46.98 of the bank's money for 12 months. To get the true interest we divide the charges, $17, by the average loan. $17 ÷ $46.81 = .362 or 36.2%.

When an automobile dealer advertises financing at 8.5% or 10%, it is a **finance rate,** not an interest rate.

EXAMPLE 2: If you buy an automobile and the balance to be financed is $8051, the **finance rate** is 10%, and the length of the loan is 3 years, how much would the payments be, and what would be the true rate of interest?

METHOD: The **finance rate** is 10%, so $8051 × .10 = $805.10. This is the amount the finance company or bank adds to the loan for *each year* of the loan, since your loan is for three years, 3 × $805.10 = $2415.30 is the amount added to the unpaid balance. This means that you will be paying the finance company:

$$\$8051 + \$2415.30 = \$10,\!466.30$$

To get the amount of the payments we divide this amount by 36, the number of months.

$$\$10,\!466.30 \div 36 = \$290.73$$

However, when we multiply $290.73 by 36 we get only $10466.28 (because we rounded off), so the monthly payments are:

35 @ $290.73 and 1 @ 290.75

To figure out the rate of interest, we must calculate the average loan; to do this, we make the following table:

| Month | Payment | Balance |
|---|---|---|
| 1 | 290.73 | 10466.30 |
| 2 | 290.73 | 10175.57 |
| 3 | 290.73 | 9884.84 |
| 4 | 290.73 | 9594.11 |
| 5 | 290.73 | 9303.38 |
| 6 | 290.73 | 9012.65 |
| 7 | 290.73 | 8721.92 |
| 8 | 290.73 | 8431.19 |
| 9 | 290.73 | 8140.46 |
| 10 | 290.73 | 7849.73 |
| 11 | 290.73 | 7559.00 |
| 12 | 290.73 | 7268.27 |
| 13 | 290.73 | 6977.54 |
| 14 | 290.73 | 6686.81 |
| 15 | 290.73 | 6396.08 |
| 16 | 290.73 | 6105.35 |
| 17 | 290.73 | 5814.62 |
| 18 | 290.73 | 5523.89 |
| 19 | 290.73 | 5233.16 |
| 20 | 290.73 | 4942.43 |
| 21 | 290.73 | 4651.70 |
| 22 | 290.73 | 4360.97 |
| 23 | 290.73 | 4070.24 |
| 24 | 290.73 | 3779.51 |
| 25 | 290.73 | 3488.78 |
| 26 | 290.73 | 3198.05 |
| 27 | 290.73 | 2907.32 |
| 28 | 290.73 | 2616.59 |
| 29 | 290.73 | 2325.86 |
| 30 | 290.73 | 2035.13 |
| 31 | 290.73 | 1744.40 |
| 32 | 290.73 | 1453.67 |
| 33 | 290.73 | 1162.94 |
| 34 | 290.73 | 872.21 |
| 35 | 290.73 | 581.48 |
| 36 | 290.75 | 290.75 |
| + | 10466.30 | 193,626.90 |

We add the column of outstanding balances to get: $193,626.92. We divide this by 36 (number of months) to get the average loan.

$$\$193,\!626.90 \div 36 = \$5378.53$$

Now to get the 3 year (36 month) interest rate, we divide the cost of the loan, $2415.30, by the average loan

$$\$2415.30 \div 5378.53 = .449 \text{ or } 45\%.$$

Now since the loan was for three years, we get the true interest rate by dividing this by 3. So the true interest is 15%.

## CREDIT CARD AND STORE CHARGE ACCOUNTS

The purchase of merchandise from a store on a charge account or using a bank credit card is usually paid back in small monthly amounts. The store or bank charges an amount based on the unpaid balance. The charge is usually a percent of the unpaid balance. Here is an example of such a transaction:

EXAMPLE 1: Mrs. Geiger purchased a TV set for $100 from the ABC appliance store. She was to pay the store monthly payments of $10.00 until the loan was repaid. The store charges 1.5% on the unpaid balance each month. What interest rate did Mrs. Geiger actually pay.

METHOD:

| Month | Amount of Loan | Payment | Interest Charge | Payment on Principal | New Balance |
|-------|---------|---------|----------|-----------|---------|
| 1 | $100.00 | $10 | $1.50 | $8.50 | $91.50 |
| 2 | $ 91.50 | $10 | $1.37 | $8.63 | $82.87 |
| 3 | $ 82.87 | $10 | $1.24 | $8.76 | $74.11 |
| 4 | $ 74.11 | $10 | $1.11 | $8.89 | $65.22 |
| 5 | $ 65.22 | $10 | $ .98 | $9.02 | $56.20 |
| 6 | $ 56.20 | $10 | $ .84 | $9.16 | $47.04 |
| 7 | $ 47.04 | $10 | $ .71 | $9.29 | $37.75 |
| 8 | $ 37.75 | $10 | $ .57 | $9.43 | $28.32 |
| 9 | $ 28.32 | $10 | $ .42 | $9.58 | $18.74 |
| 10 | $ 18.74 | $10 | $ .28 | $9.72 | $ 9.02 |
| 11 | $ 9.02 | $ 9.16 | $ .14 | $9.02 | $   0 |

A look at this table shows it took Mrs. Geiger 11 months to repay the loan. To get the average loan we add the column *Amount of Loan* and divide by 11 (the number of months).

The sum of the *Amount of Loan* column is $610.77. $610.77 ÷ 11 = $55.52 Now Mrs. Geiger payed 10 × $10 + 9.16 = $109.16, so the finance charge was $9.16. $9.16 ÷ $55.52 = .165 or 16.5% for 11 months. Now 16.5% × $\frac{12}{11}$ = 18% interest for 12 months. This is just what we would expect, since $1\frac{1}{2}$% per month is 1.5 × 12 = 18.

**Practice Exercise No. 90**

1. What would be the actual interest rate paid on $80 payable at $10 each month at 3% a month until the loan is paid? How many months will it take? What is the amount of the last payment?

2. What will be the total amount paid on an installment purchase of $200 at 2% a month on the unpaid balance if it is repaid at $30 a month. What will be the amount of the last payment?

3. Calculate the monthly payments and the true interest rate on an automobile loan. The terms of this loan are $2000.00 for 1 year at a finance rate 8.8%.

# CHAPTER TEN

# MEASUREMENT OF DISTANCE, WEIGHT AND TIME

We learn from history that many different groups of people throughout the world developed systems of measurements for their own needs. Using some methods that we would regard as crude, and others that were more accurate, they devised ways of measuring distance, weighing objects, judging the passage of time and so forth.

The Egyptians gave evidence of the earliest systems of measuring distance in the building of their pyramids, which date back to 3000 B.C. As might be expected, in the first systems for measuring distance, comparisons were made to sizes of parts of the body. You must have guessed that the 12-inch foot ruler comes from the size of a man's foot despite the fact that men's feet vary in size from 6 inches to 20 inches. Fingernails, digits, arms, hand spans all became standards of length.

A popular early standard of measuring length was the *cubit* used by the Egyptians. It was the length of a forearm from the point of the elbow to the end of the middle finger. As you can see, the size of any cubit would depend upon the size of the forearm of the man doing the measuring. It was not very standardized.

Another criterion of measurement, in colonial America, was the *hand*. It referred to the width of a man's hand with the fingers together. The heights of horses were measured in hands. It would be stated that a particular horse stood 15 hands high. If the man doing the measuring had large, wide hands, he might claim to be buying a small horse by his standards, although the horse could have been much above average height. This caused a good deal of bickering. Today, a hand equals 4 inches.

On the subject of horses, in 1500 the English mile was established as eight furlongs. If any parent wants to know, a furlong is 40 rods.

The first approach to real standardization of measurements, based on such variable parts of the body, was made by King Henry I of England. Issuing a royal decree, he announced that the distance from the point of his nose to the end of his thumb on his outstretched arm as the lawful **yard.**

To finally fix this distance or standardize it, as we say, a bronze bar of this length was kept as the Standard of Reference in the King's Exchequer in England.

In 1885 two copies of this standard were sent to the United States and later accepted by the Office of Weights and Measures as legal standards of the United States.

Today at the Office of Weights and Measures in Washington, D.C., we have extended this practice of maintaining standards for all of our units of measure such as the inch, foot, pound, and others which we shall learn to use in this chapter.

## DENOMINATE NUMBERS

In every part of our daily living we use numbers to tell us about quantities or amounts such as 3 boys, 4 cows, 5 trees. This is the application of numbers to objects that we name or describe. However, a **denominate number** is one that refers to a unit of measurement that has been established by law or general usage. 1 *quart*, 2 *inches*, 5 *pounds*, 60 *degrees* are examples of denominate numbers.

A **compound denominate number** is one that consists of two or more units of the same kind, as 1 foot 2 inches, 3 hours 10 minutes, 1 pound 3 ounces.

Denominate numbers are used to express measurements of many kinds, such as:

(a) Weight (pounds)
(b) Time (seconds)
(c) Linearity (feet)
(d) Temperature (degrees)
(e) Area (square inch)
(f) Volume (cubic yards)
(g) Angularity (degrees)
(h) Liquids (quarts)

This classification is not complete. Systems of currency (dollars and cents, pounds sterling and pence, etc.) as well as the various foreign systems of weights and measures are also denominate numbers.

To gain facility in working out arithmetic problems involving denominate numbers, it is necessary to know the most common tables of measures such as are given here for reference. Take note of the abbreviations since they are given in the manner in which the values are usually written.

## TABLES OF MEASURES

### Linear Measure—Measures Lengths or Distances

12 inches (in. or ") = 1 foot (ft. or ')
3 feet or 36 inches = 1 yard (yd.)
$5\frac{1}{2}$ yards or $16\frac{1}{2}$ feet = 1 rod (rd.)
220 yards or $\frac{1}{8}$ mile = 1 furlong (fur.)
320 rods or 8 furlongs = 1 mile (mi.)
1760 yards = 1 mile
5280 feet = 1 mile

### Weight Measures

There are four different measures of weight. The common one is the **avoirdupois.**

16 ounces (oz.) = 1 pound (lb.)
100 pounds = 1 hundredweight (cwt.)
2000 pounds = 1 ton or short ton
112 pounds = 1 cwt. old measure
2240 pounds = 1 long ton

Although the others are used for special purposes and will not be studied here, you should at least know what they are and how they are used. They are:

(a) **Troy**—for weighing gold, silver, and other precious metals.

(b) **Apothecaries'**—used by druggists for weighing chemicals.

(c) **Metric**—used in scientific work.

### Liquid Measure

2 measuring cups = 1 pint (pt.)
16 fluid ounces (fl. oz.) = 1 pint (pt.)
4 gills (gi.) = 1 pint (pt.)
2 pints $\left.\right\}$ = 1 quart (qt.)
32 fluid ounces
4 quarts = 1 gallon (gal.)
$31\frac{1}{2}$ gallons = 1 barrel (bbl.)
2 barrels = 1 hogshead (hhd.)

### Dry Measure

2 pints (pt.) = 1 quart (qt.)
8 quarts = 1 peck (pk.)
4 pecks = 1 bushel (bu.)
$2\frac{5}{8}$ bushels = 1 barrel (bbl.)

### Units of Counting

12 units = 1 dozen (doz.)
12 dozen $\left.\right\}$ = 1 gross (gr.)
144 units
24 sheets = 1 quire
480 sheets = 1 ordinary ream
500 sheets = 1 printer's ream

### Surface Measure or Square Measure

144 square inches (sq. in.) = 1 square foot (sq. ft.)
9 square feet = 1 square yard (sq. yd.)
$30\frac{1}{4}$ square yards = 1 square rod (sq. rd.)
160 square rods $\left.\right\}$ = 1 acre (A.)
43,560 square feet
640 acres = 1 square mile (sq. mi.)

## Volume Measure or Cubic Measure

1728 cubic inches = 1 cubic foot
(cu. in.)            (cu. ft.)
27 cubic feet      = 1 cubic yard
(cu. yd.)
128 cubic feet    = 1 cord of wood (cd.)

## Time Measure

60 seconds (sec.) = 1 minute (min.)
60 minutes        = 1 hour (hr.)
24 hours          = 1 day (da.)
7 days            = 1 week (wk.)
30 days           = 1 month (mo.) (for
interest, etc.)
360 days          = 1 year (yr.) (for
interest calculations)
12 months ⎫
365 days  ⎭       = 1 year (yr.)
366 days          = 1 leap year
10 years          = 1 decade
100 years         = 1 century (C.)

## Angle Measure

60 seconds (″)    = 1 minute (′)
60 minutes        = 1 degree (°)
90 degrees        = 1 right angle (∟)
360 arc degrees   = 4 right angles
360 angle degrees = 1 circumference (○)

## EQUATING DENOMINATE NUMBERS

Every measurement has many units that are related to each other. For instance, in linear measure we use inches, feet, yards, rods and miles. In any situation you would use the units that are best suited to the distances concerned. You would measure distances between cities in *miles*. You would measure the distance from the top to the bottom of this page in *inches*. Therefore it is important that you learn thoroughly the values and relationships of the most common units in the tables.

### Practice Exercise No. 91

Fill in the missing numbers. Use the preceding tables if necessary.

1. 1 ft. = (    ) inches
2. (    ) ounces = 1 lb.
3. 1 peck = (    ) qt.
4. 1 long ton = (    ) lb.
5. (    ) bushels = 1 barrel
6. (    ) pints = 1 quart
7. 1 dozen = (    ) units
8. 1 barrel = (    ) gallons
9. (    ) sheets = 1 quire
10. 1 pint = (    ) gills
11. 1 ream = (    ) sheets
12. 1 rod = (    ) yards
13. (    ) seconds = 1 min.
14. 1 week = (    ) days
15. 1 leap yr. = (    ) days
16. (    ) yards = 1 mile
17. 1 ton = (    ) hundredweight
18. 1 gross = (    ) dozen
19. 1 yd. = (    ) feet
20. (    ) fluid ounces = 1 pint
21. 1 bushel = (    ) pecks
22. 1 printer's ream = (    ) sheets
23. 1 mile = (    ) rods
24. (    ) units = 1 gross
25. 1 hour = (    ) minutes
26. (    ) in. = 1 yd.
27. (    ) days = 1 week
28. (    ) lb. = 1 ton
29. 1 day = (    ) hours
30. 1 gal. = (    ) quarts
31. 1 year = (    ) days
32. (    ) ft. = 1 rod
33. (    ) lb. = 1 cwt.
34. 1 qt. = (    ) fluid ounces
35. 1 mile = (    ) feet
36. 1 decade = (    ) years

Name some common object that is measured in the following units:

| | | |
|---|---|---|
| **37.** ounces | **42.** gallon | **47.** miles |
| **38.** feet | **43.** quart | **48.** barrel |
| **39.** inches | **44.** quire | **49.** ton |
| **40.** bushel | **45.** century | **50.** dozen |
| **41.** gill | **46.** month | |

## APPLYING THE FOUR FUNDAMENTAL OPERATIONS WITH DENOMINATE NUMBERS

### Addition of Denominate Numbers

In adding denominate numbers, the procedure is the same as was used in learning fundamental addition with place values, except that the *carrying* or *regrouping* must fit the measure.

EXAMPLE 1: Add 2 ft. 8 in. and 1 ft. 6 in.

METHOD:

```
        1 carry
        2 ft.     8 in.
    +   1 ft.     6 in.
        4 ft.     2 in. ANS.
```

EXPLANATION: List like units under each other. Add the inches. Change 14 inches to 1 ft. 2 in.

Put down the 2 inches, carry the 1 to the foot column and add this column.

EXAMPLE 2: Add 4 gal. 3 qt. 1 pt. and 2 gal. 2 qt. 1 pt.

METHOD:

```
        1 carry   1 carry
        4 gal.    3 qt.     1 pt.
    +   2 gal.    2 qt.     1 pt.
        7 gal.    2 qt.     0 pt. ANS.
```

EXPLANATION: List like units under each other. Add the pints. Since 2 pints = 1 quart, we have zero (0) pt. and carry 1 qt. to the next column. Add the quart column which is 6 qt. Change to 1 gal. 2 qt. Put

down 2 qt. and carry 1 gal. to the next column. Adding gives 7 gal.

### Practice Exercise No. 92

Add the denominate numbers which follow.

| | |
|---|---|
| **1.** 4 ft. 7 in.<br>   3 ft. 7 in. | **6.** 6 bu. 2 pk.<br>   5 bu. 1 pk. |
| **2.** 23 lb. 12 oz.<br>   4 lb.  6 oz. | **7.** 5 lb. 9 oz.<br>   2 lb. 6 oz. |
| **3.**  4 qt. 1 pt.<br>  12 qt. 1 pt. | **8.** 3 pt. 12 fl. oz.<br>   1 pt.  8 fl. oz. |
| **4.** 4 hr. 20 min.<br>   2 hr. 45 min. | **9.** 15 ft. 10 in.<br>    3 ft.  9 in. |
| **5.** 2 yd.     9 in.<br>  1 yd. 1 ft. 5 in. | **10.** 6 hr. 20 min. 10 sec.<br>   7 hr. 40 min. 35 sec.<br>   2 hr. 12 min. 40 sec. |

### Subtraction of Denominate Numbers

In the subtraction of denominate numbers we use the method of *exchange* or borrowing in the same way as for ordinary numbers, except that the exchange must fit the measure.

EXAMPLE 1: From 23 lb. 4 oz. take 4 lb. 8 oz.

METHOD:

$$23 \text{ lb. } 4 \text{ oz.} = 22 \text{ lb. } 20 \text{ oz.}$$
$$- \ 4 \text{ lb. } 8 \text{ oz.} = \ \underline{4 \text{ lb. } \ 8 \text{ oz.}}$$
$$18 \text{ lb. } 12 \text{ oz. ANS.}$$

EXPLANATION: Since we can't take 8 oz. from 4 oz., we exchange 1 lb. for 16 oz. and add it to the 4 oz. to make 20 oz. Then subtract as usual.

EXAMPLE 2: From 5 ft. 4 in. take 2 ft. 8 in.

METHOD:

```
        4 ft.    16 in.
        5 ft.     4 in.
    −   2 ft.     8 in.
        2 ft.     8 in. ANS.
```

EXPLANATION: Exchange 1 ft. for 12 in. and add it to the 4 in. Then subtract as usual.

NOTE: In the example the exchanges are shown, but you should learn to make them without noting them on paper.

### Practice Exercise No. 93

Subtract and check by addition.

| | | | |
|---|---|---|---|
| **1.** | 10 wk. 5 da.<br>− 4 wk. 6 da. | **6.** | 3 mi.<br>− 1 mi. 880 yd. |
| **2.** | 12 yr. 3 mo.<br>− 8 yr. 5 mo. | **7.** | 8 lb. 2 oz.<br>− 6 lb. 8 oz. |
| **3.** | 9 ft. 6 in.<br>− 4 ft. 9 in. | **8.** | 12 yd. 1 ft.<br>− 6 yd. 2 ft. |
| **4.** | 5 min. 12 sec.<br>− 3 min. 44 sec. | **9.** | 10 gal. 1 qt.<br>− 7 gal. 2 qt. |
| **5.** | 8 tons<br>− 3 tons 800 lb. | **10.** | 5 lb. 7 oz.<br>− 1 lb. 12 oz. |

## Multiplication of Denominate Numbers

Multiplication is a fast way of adding. Let's examine what takes place when we multiply 3 ft. 7 in. by 4.

EXAMPLE 1:

*By addition:*

$$\begin{array}{r} 3 \text{ ft.} \quad 7 \text{ in.} \\ 3 \text{ ft.} \quad 7 \text{ in.} \\ 3 \text{ ft.} \quad 7 \text{ in.} \\ \underline{3 \text{ ft.} \quad 7 \text{ in.}} \\ 12 \text{ ft. 28 in.} = \\ 14 \text{ ft.} \quad 4 \text{ in. ANS.} \end{array}$$

EXPLANATION: Add the inches, change the 28 in. into 2 ft. 4 in. because 12 in. = 1 ft. Put down the 4 in. and add the 2 ft. to the 12 ft.

*By multiplication:*

$$\begin{array}{r} 3 \text{ ft.} \quad 7 \text{ in.} \\ \underline{\times 4} \\ 12 \text{ ft. 28 in.} = \\ 14 \text{ ft.} \quad 4 \text{ in. ANS.} \end{array}$$

EXPLANATION: (1) Multiply each of the parts of the measure by the multiplier. (2) Simplify the units starting with the lower values on the right.

EXAMPLE 2: Multiply 3 hr. 12 min. by 6.

METHOD:

$$\begin{array}{r} 3 \text{ hr. 12 min.} \\ \underline{\times 6} \\ 18 \text{ hr. 72 min.} = \\ 19 \text{ hr. 12 min. ANS.} \end{array}$$

EXPLANATION: (1) Multiply each measure by the multiplier. (2) Simplify.

### Practice Exercise No. 94

Multiply the denominate numbers which follow.

| | | | |
|---|---|---|---|
| **1.** 2 hr. 35 min.<br>× 3 | | **6.** 1 ton 900 lb.<br>× 4 | |
| **2.** 2 gal. 3 qt.<br>× 4 | | **7.** 2 mi. 880 yd.<br>× 3 | |
| **3.** 3 lb. 9 oz.<br>× 2 | | **8.** 2 hr. 25 min.<br>× 3 | |
| **4.** 4 yd. 2 ft.<br>× 7 | | **9.** 6 bu. 2 pk.<br>× 5 | |
| **5.** 8 qt. 1 pt.<br>× 3 | | **10.** 2 qt. 20 fl. oz.<br>× 3 | |

## Division of Denominate Numbers

In proceeding with the division of denominate numbers, divide each part of the denominate number beginning with the largest unit. If there is a remainder to any part, it is *changed and added* to the next part.

EXAMPLE 1: Divide 13 hr. 30 min. by 3.

METHOD:

$$\begin{array}{r} 4 \text{ hr. 30 min. ANS.} \\ 3 \overline{)\; 13 \text{ hr. 30 min.}} \\ \underline{12 \text{ hr.}} \\ 1 \; + \; 30 = 90 \text{ min.} \end{array}$$

EXPLANATION: 13 hr. ÷ 3 = 4 with 1 hr. remainder. Change 1 hr. to 60 min. and add to 30 min. 90 ÷ 3 = 30 min.

EXAMPLE 2: Divide 38 lb. 10 oz. by 4.

METHOD:

$$\begin{array}{r} 9 \text{ lb. } 10\frac{1}{2} \text{ oz. ANS.} \\ 4\overline{)\ 38 \text{ lb. } 10 \text{ oz.}} \\ \underline{36} \\ 2 \text{ lb. } + 10 \text{ oz. } = 42 \text{ oz.} \end{array}$$

EXPLANATION:
(1) Divide into pounds.
(2) Change remainder to ounces and add.
(3) Divide into ounces, change remainder into fraction.

### Practice Exercise No. 95

Divide the denominate numbers which follow.

1. $2\overline{)\ 3 \text{ hr. } 6 \text{ min.}}$       6. $5\overline{)\ 42 \text{ min. } 30 \text{ sec.}}$

2. $4\overline{)\ 10 \text{ yd. } 8 \text{ in.}}$       7. $8\overline{)\ 12 \text{ qt. } 1 \text{ pt.}}$

3. $3\overline{)\ 11 \text{ ft. } 5 \text{ in.}}$       8. $3\overline{)\ 4 \text{ qt. } 12 \text{ fl. oz.}}$

4. $2\overline{)\ 5 \text{ pt. } 11 \text{ fl. oz.}}$       9. $4\overline{)\ 11 \text{ lb. } 2 \text{ oz.}}$

5. $6\overline{)\ 12 \text{ bu. } 6 \text{ pk.}}$       10. $2\overline{)\ 1 \text{ yd. } 26 \text{ in.}}$

### Practice Exercise No. 96

The following problems will test your skill in working with denominate numbers.

**1.** Joan did babysitting for her neighbor at $2.75 an hour. She sat from 7:30 P.M. to 11:50 P.M. How much did she earn?

**2.** Mrs. Abernathy bought two squash. One weighed 3 lb. 12 oz. and the other weighed 4 lb. 10 oz. At 30¢ per lb., how much did she have to pay for both squash?

**3.** Dick bought 5 gallons of paint for his porch. When he finished he had 1 gallon 3 quarts left. How much paint did he use?

**4.** Norm worked on his science project for 2 hours 20 minutes on Monday. On Tuesday he worked for 1 hour 45 minutes. On Thursday he spent 35 minutes on it. How much time did he spend on the project?

**5.** The afternoon train is due to arrive at the Newark station at 2:35 P.M. The station master reports it will be 1 hour 50 minutes late. What time will it arrive?

**6.** Jim caught five fish varying in weight from 1 pound to 4 pounds. Altogether the fish weighed 12 pounds 8 ounces. What is the average weight per fish?

**7.** Sheila decided to make curtains for four equal size windows. How much length can she allow for each window if she has 10 yards 8 inches of cloth?

**8.** A vegetable dealer bought eight bushels of fresh string beans. Each weighed 8 lb. 10 oz. What should be the total weight of the delivery?

**9.** John, Bob and Joe went berry picking and agreed to share equally all they picked. When they finished they had a total of 2 pecks and 2 quarts. How much did each one receive?

**10.** Mr. Bimler calculated that it took on the average 3 hours and 15 minutes to complete a custom-made suit in his shop. How many eight-hour work days must he allow to complete two dozen such suits?

### Changing from Higher to Lower Units of Denominate Numbers

In changing from yards to feet, quarts to pints or pounds to ounces you change from higher value units to lower value units of denominate numbers.

**Rule: To change from higher to lower denominate number units,** multiply *the higher unit by the proper equivalent factor.*

EXAMPLE 1: Change $4\frac{1}{2}$ yards to inches.

METHOD: Think, ? inches = 1 yard.
          36 in. = 1 yd.

Therefore, $4\frac{1}{2} \times 36 =$

$$\frac{9}{2} \times \frac{\overset{18}{\cancel{36}}}{\underset{1}{1}} = 162 \text{ in. ANS.}$$

EXAMPLE 2: Change $6\frac{1}{2}$ gallons to pints.

METHOD: Think, ? pints = 1 gallon.
          8 pt. = 1 gal.

Therefore, $6\frac{1}{2} \times 8 =$

$$\frac{13}{2} \times \overset{4}{\underset{1}{8}} = 52 \text{ pt. ANS.}$$

**Rule: To change from lower to higher denominate number units,** divide *the lower unit by the proper equivalent factor.*

EXAMPLE 1: How many tons is an automobile weighing 3600 lb.?

METHOD: Think, $\underline{?}$ pounds = 1 ton.
2000 lb. = 1 ton

Therefore, $\dfrac{3600}{2000} = \dfrac{36}{20} = \dfrac{9}{5} = 1\frac{4}{5}$ ton ANS.

### Practice Exercise No. 97

Fill in the blanks below.

1. 2 ft. 8 in. = _____ in.

2. 5 qt. 20 fl. oz. = _____ fl. oz.

3. 85 min. = _____ hr. _____ min.

4. 29 pt. = _____ gal. _____ pt.

5. 40 oz. = _____ lb.

6. 3 gal. 1 pt. = _____ pt.

7. $4\frac{1}{3}$ yd. = _____ ft.

8. 14 pk. = _____ bu. _____ pk.

9. $\frac{2}{3}$ yd. = _____ in.

10. $\frac{1}{5}$ hr. = _____ min.

11. 6 qt. 1 pt. = _____ qt.

12. _____ oz. = $\frac{7}{8}$ lb.

13. 3 pt. 8 fl. oz. = _____ pt.

14. 4 bu. 3 pk. = _____ bu.

15. _____ lb. = 10 oz.

16. $\frac{1}{4}$ pk. = _____ qt.

17. 8 rd. = _____ ft.

18. 1 ft. 3 in. = _____ ft.

19. 6 yd. 2 ft. = _____ yd.

20. 96 in. = _____ yd.

21. 34 oz. = _____ lb.

22. 110 min. = _____ hr.

23. _____ oz. = $\frac{1}{8}$ lb.

24. _____ ton = 500 lbs.

25. $\frac{1}{10}$ mile = _____ ft.

26. 48 fl. oz. = _____ qt. _____ fl. oz.

27. 24 in. = _____ yd.

28. 48 in. = _____ ft.

29. _____ qts. = 10 gal.

30. $\frac{1}{2}$ pt. = _____ fl. oz.

### THE METRIC SYSTEM OF MEASUREMENTS

In the beginning of our study of denominate numbers, we dealt with various units of measure based on the early English system of measurements and modifications of the practices of many ancient people.

Coming closer to modern times, in France and in other parts of the world as well as in the United States, a different system of measurements has been introduced. It is called the **metric** system. This system was specifically designed for convenience and efficiency of use rather than as an outgrowth of ancient practices based on the sizes of parts of the body.

The metric system was devised in 1799 in France. The French government had engineers calculate how far it was from the North Pole to the Equator and then took one ten-millionth of this length and called it a *meter*. By design, it is a *decimal* system in which the key units of measurement are related to each other by multiples of 10. In the metric system many computations can be easily performed just by moving the decimal point to the left or right.

Another advantage of the metric system is that the measures of length, volume and weight are *related* to each other.

## LINEAR MEASURE IN THE METRIC SYSTEM

As noted, the **meter** (M.) is the prime unit of length in the metric system. A meter, which is a little longer than our yard, is 39.37 inches.

The meter is further divided into ten parts or **decimeters** (Figure 12).

Each decimeter is further divided into ten parts called **centimeters.** There are 100 centimeters in a meter (Figure 13).

Each centimeter is further divided into ten parts called **millimeters.** There are 1000 millimeters in a meter.

It may help you to memorize these measures if you recognize that the Latin prefixes offer a valuable aid in understanding them. For example, *deci* means 10, *centi* means 100 and *milli* means $\frac{1}{1000}$ (not 1000). The prefix for 1000 in this system is *kilo*.

Examine the section of a ruler which has both scales on it. You may observe some relationships which shall be pointed out in the sections which follow (Figure 14).

With the information given thus far, you may note that to change 584 meters to centimeters, you would merely add two zeros and get 58,400 centimeters.

To change 763 centimeters to meters, you need only move the decimal point 2 places to the left to give 7.63 meters.

Compare this last computation with our system. To change 763 inches to yards, you have to divide by 36. Of the two processes, which appears easier?

In the larger units the metric system uses the **kilometer** which is equal to 1000 meters. This is the closest measure to our mile and is about $\frac{5}{8}$ of a mile. In foreign countries the maps are marked in kilometers and you will now find the speedometers of cars calibrated in "kilometers per hour" as well as "miles per hour."

FIGURE 12  One decimeter.

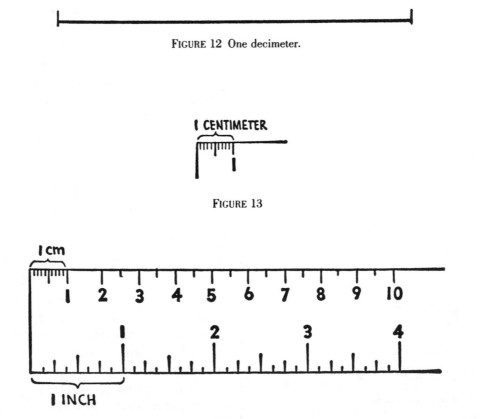

FIGURE 13

FIGURE 14

## METRIC TABLES OF MEASURES

Decimal Relationships in the Metric System

| | Unit | | Meters | | U.S. Value |
|---|---|---|---|---|---|
| | 1 millimeter (mm.) | = | .001 | = | .03937 in. |
| 10 millimeters | = 1 centimeter (cm.) | = | .01 | = | .3937 in. |
| 10 centimeters | = 1 decimeter (dm.) | = | .1 | = | 3.937 in. |
| 10 decimeters | = 1 meter (M.) | = | 1. | = | 39.37 in. |
| 10 meters | = 1 dekameter (Dm.) | = | 10. | = | 32.809 ft. |
| 10 dekameters | = 1 hectometer (Hm.) | = | 100. | = | 328.09 ft. |
| 10 hectometers | = 1 kilometer (Km.) | = | 1000. | = | .62137 mile |

The most commonly used linear metric measures and English equivalents are usually shown this way:

10 millimeters = 1 centimeter = .3937 inch
100 centimeters = 1 meter = 39.37 inches
1000 meters = 1 kilometer = $.621^+$ or $\frac{5}{8}$ mile
39.37 inches = 1 meter; 1 inch = 2.54 centimeters

### Practice Exercise No. 98

Answer the questions and solve the problems which follow.

1. An inch = ____ centimeters

2. A decimeter = ____ inches

3. 3 centimeters = ____ millimeters

4. An inch = ____ millimeters

5. 3 mm. = $\frac{?}{10}$ cm.

6. 2.3 Km. = ____ M.

7. 4 M. = ____ cm.

8. 4825 M. = ____ Km.

9. 25 cm. = ____ mm.

10. 12 M. = ____ cm.

11. 3.41 Km. = ____ M.

12. .007 M. = ____ mm.

13. 257 cm. = ____ M.

14. 5823 mm. = ____ cm.

15. 5823 mm. = ____ M.

16. 582.3 cm. = ____ M.

17. 5 M. = ____ Km.

18. .8 Km. = ____ M.

19. .09 Km. = ____ M.

20. .004 Km. = ____ M.

21. How many yards are there in the 100 meter dash?

22. How many yards are in the 400 meter run?

23. Which is the greater distance, the mile run or the 1500 meter run; and by how much distance do they differ, expressed in yards to the nearest whole yard?

24. If you are 5 ft. 6 in. tall, what would your height be in centimeters?

25. How many inches wide is a 35 mm. film?

26. The distance from Paris to London is 336 kilometers. How many miles is it?

27. The speed limit is 50 mi. per hr. What is this equal to in kilometers per hour on the speedometer of the foreign car?

### Measuring Weights by the Metric System

In the metric system, the unit of weights is the **gram.**

The **gram** (*gm.*) is equal to the weight of a cube of water that is 1 cm. on an edge (Figure 15). It is equal to .035 oz. in our units of weight (avoirdupois).

1 cu. cm of water
1 gram weight

FIGURE 15. 1 cu. cm. of water equals 1 gm. weight.

Weights in grams and fractions are used in measuring out dosages of medicines. This system is used extensively by pharmacists. The *centi*gram is .01 gram and the *milli*gram equals .001 gram.

Larger metric-weight units are used by the housewife and in industry in Europe. The kilogram (kg.), which equals 1000 grams, is used more than any other unit of weight and is equal to 2.2 pounds.

In our system of weights, 1 lb. = 454 gms (approx.) and 1 oz. = 28.4 gms (approx.).

### Practice Exercise No. 99

Solve the following problems.

**1.** If Peter weighs 140 lb., how many kilograms would he weigh in the French metric system?

**2.** If you were in a foreign country where the metric system is used, what would you ask for if you wanted *approximately:* (a) 1 lb. of butter, (b) 2 lb. of apples, (c) 1 oz. of cinnamon, (d) 5 lb. of potatoes?

**3.** A one-half oz. letter costs 7¢ to airmail in some foreign countries. What is the weight in grams?

**4.** In a 2 oz. package, how many pills will you get if each pill weighs 200 milligrams?

**5.** A foreign airline allows a passenger to take aboard 25 kg. of luggage without extra charge. Approximately how many pounds is that?

## FIGURING CLOCK TIME IN DIFFERENT PARTS OF THE COUNTRY

Clock time is related to positions of the sun. It is also referred to as solar time. We

FIGURE 16

say that the sun rises in the east and sets in the west. We know that it is the earth that rotates.

We say that it is 12 o'clock M (12 noon) when the sun is directly overhead. This is at the place when the sun crosses its meridian. Thus, if it is noon where you are, it is **afternoon** (P.M. or post meridian) at all places **east** of you, because the sun had been overhead in these places before it reached you. To the **west** of you, it is forenoon (A.M. or ante meridian), because the sun is not yet overhead in these places.

Because the earth turns 360 degrees in 24 hours, places that are 15 degrees ($\frac{1}{24}$ of 360°) to the east of you, should be one hour ahead of your time. Places 15° to the west of you, should be one hour earlier (behind) than your time.

If you take a plane trip across the United States you must set your watch ahead or back at different locations to obtain the proper time.

If you start from New York and travel *west* to San Francisco, you would set your watch *back* an hour, three separate times.

The map (Figure 16) will show you that the United States is divided into four time zones, each approximately 15° of longitude in width. The four time zones are Eastern, Central, Mountain, and Pacific.

Every place in each of these zones uses the same *standard time*. For example, in the Eastern time zone, New York, Cleveland and Miami have the same time.

*A time zone to the east is one hour ahead of its neighboring zone to the west.*

For example, New York in the Eastern zone is one hour ahead of Chicago in the Central zone.

*A time zone to the west is one hour earlier (behind) than its neighboring zone to the east.*

As an example, Denver in the Mountain zone is one hour earlier than Chicago in the Central zone.

### How to Figure Travel and Other Time Differences

EXAMPLE 1: A plane left New York at 2:30 P.M. for Chicago. It arrived there non-stop, 4 hours 40 minutes later. What time was it in Chicago?

METHOD:

1. 2:30 P.M., E.S.T. = 1:30 C.S.T.
2. 　1:30
　+ 4:40
　5:70 = 6:10 P.M.

*Step 1.* Change the starting time to the time it would be in the zone where the trip ends.
*Step 2.* Add the time of the trip.
If a trip starts and ends in the same time zone, omit Step 1.

EXAMPLE 2: A plane left Chicago at 3:30 A.M., C.S.T. and arrived in New York at 6:50 A.M., E.S.T. Find the total time in the air?

METHOD:

1. 3:30 A.M., C.S.T. = 4:30 A.M., E.S.T.
2. 　6:50
　− 4:30
　2 hr. 20 min.

*Step 1.* Convert time of leaving to time it would be in zone where the trip ends.
*Step 2.* Find the difference.

### Practice Exercise No. 100

Use the map (Figure 16) for help if needed.

**1.** When it is 4 P.M. in Salt Lake City, what time is it in (a) New Orleans, (b) Washington, D.C., (c) Seattle, (d) San Francisco, (e) Denver?

**2.** You set your watch in New York on standard time and take a plane to Los Angeles. To be in keeping with Los Angeles time, will you set the time ahead or back and how much?

**3.** If Denver is 2 hours behind New Jersey time, approximately how many degrees of longitude is there between Denver and New Jersey?

**4.** The radio announcer states that the President will speak on a nationwide hook-up at 10:30 P.M., Eastern Standard Time. What time must people in Seattle, Washington tune in?

**5.** The Rose Bowl game starts at 2 P.M., Pacific Standard Time. If you are in Delaware what time will you tune in on your television set to see the kickoff?

**6.** A Chicago to San Francisco plane leaves Chicago at 9:30 A.M., Central Standard Time. The trip takes seven hours and fifteen minutes. Using Pacific Standard Time, what time will the plane arrive in San Francisco?

**7.** A New York to Miami plane leaves New York at 10:15 A.M., Eastern Standard Time. The trip takes four hours and thirty minutes. Using Eastern Standard Time, what time will the plane arrive in Miami?

**8.** A Los Angeles to Denver plane leaves Los Angeles at 9:30 A.M., Pacific Standard Time. The trip takes two hours and twenty minutes. Using Mountain Standard Time, what time will the plane arrive in Denver?

**9.** A New Orleans to New York plane leaves New Orleans at 11:15 A.M., Central Standard Time. The trip takes five hours and forty minutes. Using Eastern Standard Time, what time will the plane arrive in New York?

**10.** A Dallas to Chicago plane leaves Dallas at 10:40 A.M., Central Standard Time. The trip takes six hours. Using Central Standard Time, what time will the plane arrive in Chicago?

## ARMED SERVICES TIME— THE 24-HOUR CLOCK

There is frequently some difficulty in figuring time with the A.M. and P.M. designations. A more convenient type of reckoning, the 24-hour clock, is becoming popular. For many years now, the armed services of the United States have kept time by the 24-hour clock.

In working the 24-hour clock, time is indicated as a four-place number, and obviously there is no need to indicate A.M. or P.M.

With the 24-hour clock, we start at midnight which is the "zero hour" or 0000. 1 A.M. is 0100 and 2 A.M. is 0200. Each hour advances the time by 100, thus 3 A.M. is 0300 and 8 A.M. is 0800. Noon becomes 1200; 1 P.M. is 1300; 11 P.M. is 2300. Minutes are indicated by units—preceded by a zero (0) if less than 10, because there must always be four digits. 8:10 A.M. is 0810; 1:15 P.M. is 1315; 11:59 P.M. is 2359; 12:01 A.M. is 0001.

You will find that after you learn to use the 24-hour clock, it becomes very much easier to calculate travel time.

### Practice Exercise No. 101

Solve these problems relating to time.

**1.** Change to time on a 24-hour clock: (a) 12:35 A.M., (b) 12:35 P.M., (c) 3 A.M., (d) 8:28 P.M., (e) 11:35 A.M.

**2.** Change to the 12-hour clock—note A.M. and P.M.: (a) 0045, (b) 1755, (c) 1203, (d) 1950, (e) 0435.

**3.** A plane took off at 0305 E.S.T. and landed at 1455 E.S.T. How long was it in the air?

**4.** A train left Chicago at 0135 C.S.T. and traveled east to New York. It arrived 19 hours and 40 minutes later. What time did the train arrive in New York, E.S.T., on the 24-hour clock?

# CHAPTER ELEVEN

# MEASURES OF LINES, ANGLES, AND PERIMETERS OF PLANE FIGURES

## LINES

Everything you look at has *lines*. They are so much a part of the objects we see, that usually we fail to observe the lines. In fact we are accustomed to recognizing objects and even faces by what we call the outlines.

The outline of any object is nothing more than a combination of the various types of lines out of which it is shaped. We see in these outlines many types of lines, let us call them by name.

### The Language of Lines

A **straight** line is the shortest distance between two points. These are all *straight* lines (Figure 17).

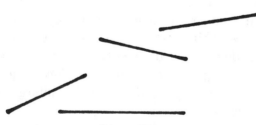

FIGURE 17

These are **curved** lines (Figure 18).

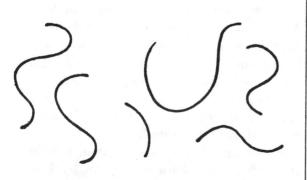

FIGURE 18

These are **broken** lines (Figure 19).

FIGURE 19

These lines are **parallel** to each other. They are the same distance apart at all points and no matter how far you continue them, they will never cross each other (Figure 20).

FIGURE 20

A **horizontal line** is a straight line that is level with the horizon (Figure 21).

FIGURE 21

A **vertical** line is a straight line that is perpendicular to the horizon (Figure 22).

FIGURE 22

A line is **perpendicular** to another line when it inclines no more to one side than the other. Such lines are said to be at right angles to each other (Figure 23).

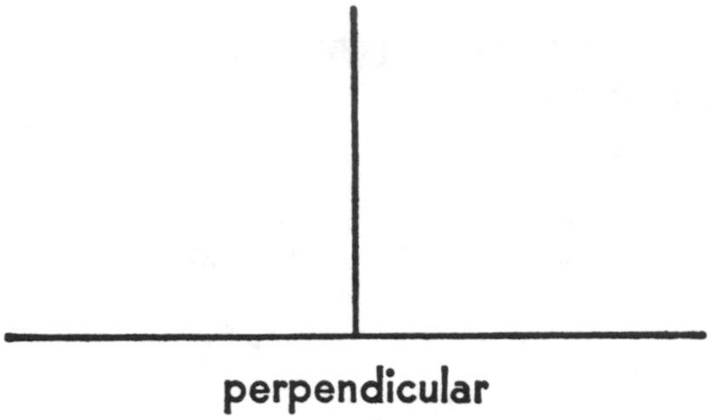

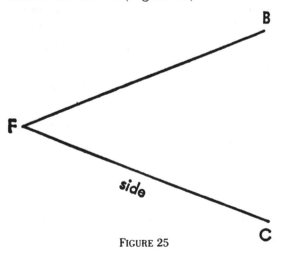

## perpendicular

FIGURE 23

An oblique line is a straight line that is neither horizontal nor vertical (Figure 24).

FIGURE 24

### ANGLES

An angle is the figure formed when two straight lines touch at a common point called the **vertex** (Figure 25).

FIGURE 25

The lines that form the angle are called the **sides.** If three letters are used to designate an angle, the vertex is read between the others. Thus Figure 25 is written ∠BFC and is read *angle* BFC: the *sides* are BF and FC.

### *Types and Sizes of Angles*

To know what an angle is, you must keep in mind that it is really composed of the spokes or radii coming from a point of focus (the vertex) which represents the center of a circle. As shown, there are 360 degrees around the point or center of the circle (Figure 26).

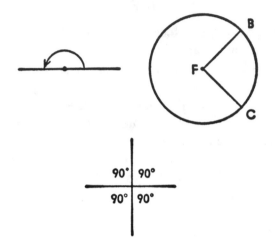

FIGURE 26

When we measured lines we used a linear measure, like inches, feet and yards. The unit of measure for angles is the *degree*. This is the symbol for a degree (°). One degree would be written 1° and it represents $\frac{1}{360}$ of a complete *revolution* around the circle or $\frac{1}{360}$th part of the circumference of a circle.

A **straight angle** represents half the distance around the circle, or 180°.

A **right angle** represents one-fourth of a revolution around the circle or 90°. Another definition for a right angle is to state that when two lines meet in a way as to form a square corner, the result is a right angle. How many right angles in a circle (Figure 27)?

An **acute angle** is less than a right angle or less than 90° (Figure 27).

An **obtuse angle** is more than a right angle, but less than a straight angle. It is thus between 90° and 180° (Figure 27).

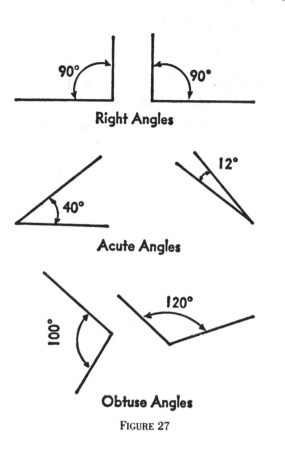

**Right Angles**

**Acute Angles**

**Obtuse Angles**

FIGURE 27

### Measuring Angles

Angles are measured by determining the part of a circle that the sides intersect. Therefore to determine the size of an angle, you measure the opening between the sides of the angle and *not* the length of the sides. To measure or "lay off" angles, a **protractor** as shown in the illustration (Figure 28) is used.

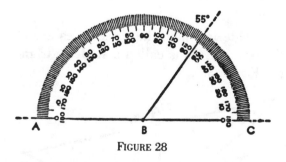

FIGURE 28

*In using the protractor, observe these features:*

(a) The diameter of the protractor is a straight line or a straight angle of 180° from

*A* to *C* as shown above. By definition, a **diameter** is a line drawn through the middle of any circle which extends to the circumference at opposite ends.

(b) The center of the protractor is at point *B* which is also the center of the circle, of which the protractor is one half.

**To measure an angle with a protractor:** Place the center of the protractor at the vertex (*B*) of the angle, and the diameter of the protractor on a line with one side of the angle (in this case, line *BC*). Read the degrees where the other side of the angle (*BD*) crosses the scale of the protractor. Is angle *CBD* above 55°?

Try measuring angle *EFG* pictured here (Figure 29). What size do you find it to be? Is it larger than ∠ *CBD* above? It is 5° greater.

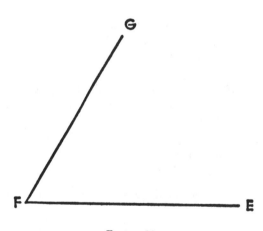

FIGURE 29

**To draw an angle of given size with a protractor:** Draw a straight line for one side of the angle. Place the center of the protractor at the point of the line that is to be the vertex of the angle, and make the straight side of the protractor coincide with the line. Place a dot on your paper at the point on the scale of the protractor that corresponds to the size of the angle to be drawn. Connect this dot and the vertex to obtain the desired angle.

### Practice Exercise No. 102

Answer the following questions pertaining to angles.

**1.** The hands of a clock or watch form angles. How many degrees are in the angle when it is 12:15 (Figure 30)?

FIGURE 30

**2.** What degree angle would be formed if it were exactly $7\frac{1}{2}$ minutes after 12?

**3.** What degree angle would be formed if it were 12:30?

**4.** If the minute hand is on the 4 and the hour hand forms an angle of 115° with the minute hand, what time is it?

**5.** Tell whether each of the following angles is acute, obtuse, a right angle or a straight angle. (a) 45°, (b) 81°, (c) 90°, (d) 180°, (e) 92°.

Use the diagram (Figure 31) for the following problems.

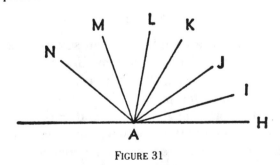

FIGURE 31

**6.** Measure ∠ *HAI*

**7.** Measure ∠ *HAJ*

**8.** Measure ∠ *HAL*

**9.** Measure ∠ *KAM*

**10.** Measure ∠ *JAN*

## TRIANGLES—FIGURES WITH THREE SIDES

A triangle is a plane figure with three sides that are joined to form three angles.

We use the sides as well as the angles as a basis for naming triangles.

The symbol for designating a triangle is △.

### *Parts of a Triangle*

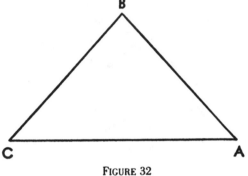

FIGURE 32

The illustrated triangle (Figure 32) would be referred to as triangle *ABC* and written as △ *ABC*.

The sides are *AB*, *BC*, and *CA*.

The angles are angle *CAB*, written ∠ *CAB*, angle *ABC* or ∠ *ABC*, and angle *BCA* or ∠ *BCA*.

### *Using the Sides to Name the Triangle*

A triangle in which the three sides are of different lengths is called a **scalene** triangle (Figure 33).

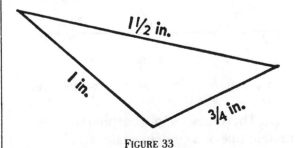

FIGURE 33

A triangle in which two sides are equal in length is called an **isosceles** triangle. In the word "isosceles" *iso* means equal and *sceles* means sides. The size of the angles that are *opposite* the equal sides in an isosceles triangle are also equal (Figure 34).

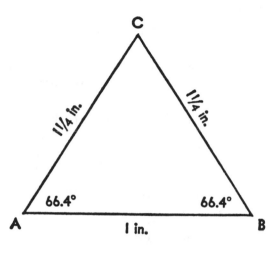

FIGURE 34

The sides *BC* and *AC* are equal.
The angles *CBA* and *CAB* are equal.

A triangle in which the three sides are equal is called an **equilateral** triangle (Figure 35). In such a triangle, the angles are *also equal*. The name **equiangular** is also applied to this type of triangle. Each angle of any equilateral triangle measures 60 degrees.

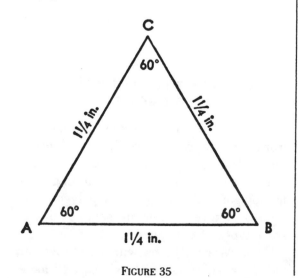

FIGURE 35

The sides:
$$AC = CB = AB$$
The angles:
$$\angle CAB = \angle ABC = \angle BCA$$

### Using the Angles to Name the Triangle

If one of the angles of a triangle is a right angle (90°), it is called a **right triangle** (Figure 36). Triangle *DEF* is a right triangle in which $\angle FDE$ is a right angle (90°). Notice how we draw a square at vertex *D* to indicate a right angle.

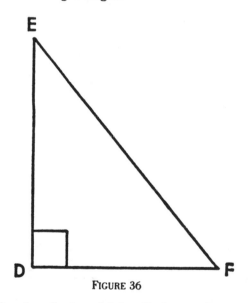

FIGURE 36

A triangle in which all the angles are acute (less than 90°) is called an **acute triangle** (Figure 37). Triangle *GHI* is an acute triangle because $\angle IGH$, $\angle GHI$, and $\angle HIG$ are each less than 90°.

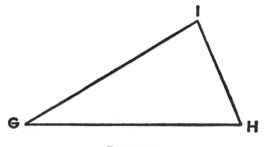

FIGURE 37

A triangle in which one angle is obtuse (more than 90°), is called an **obtuse trian-**

**gle** (Figure 38). In the obtuse triangle *UKL*, ∠ *KLU* is more than 90°.

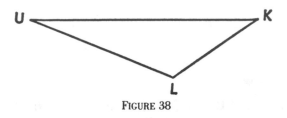

FIGURE 38

We have learned that in an equilateral or equiangular triangle each of the angles was 60°. The sum of these is therefore 180°. If you check the angles of any other triangle, you will observe the fact that:

**The sum of the angles of any triangle is 180°**

*Try this:* 1. Draw a triangle on a piece of paper. Make it any type you choose, either a right triangle, acute, obtuse, isosceles, equilateral or scalene. Label the vertexes as shown (Figure 39).

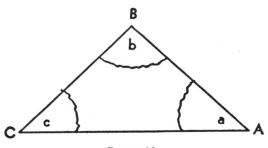

FIGURE 39

2. Cut out the triangle and tear off the three corners, keeping each vertex as an angle.

3. Place the torn pieces next to each other, with the torn part facing out as in the illustration (Figure 40).

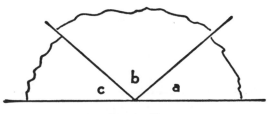

FIGURE 40

Do your three pieces make a straight angle of 180°? They should.

EXAMPLE 1: If you know that two angles of a triangle are 72° and 53° how many degrees is the third angle? What two names may be applied to such a triangle?

METHOD: 1. We know that the sum of the angles of a triangle is 180°.

2. We know that two angles equal 72° + 53° or 125°.

3. Therefore the third angle is 180° − 125° or 55°.

4. Since all three angles are less than 90° and none are equal, it is an *acute triangle* and a *scalene triangle*.

Can we have an acute triangle that is not a scalene triangle?

What type might it be?

**Practice Exercise No. 103**

Find the missing angle in the following triangles and give two names to the triangles.

| | ∠A | ∠B | ∠C | Name in terms of sides | Name in terms of angles |
|---|---|---|---|---|---|
| **1.** | 46° | 58° | ? | ? | ? |
| **2.** | 90° | 60° | ? | ? | ? |
| **3.** | 45° | 45° | ? | ? | ? |
| **4.** | 110° | 40° | ? | ? | ? |
| **5.** | 75° | 63° | ? | ? | ? |
| **6.** | 80° | 50° | ? | ? | ? |
| **7.** | 60° | 60° | ? | ? | ? |
| **8.** | 28° | 36° | ? | ? | ? |

## FIGURES WITH FOUR SIDES OR MORE

**Polygons** are many sided figures. The word "poly," pronounced "pol-lee," means many. Therefore all plane figures with more than two sides can be referred to in general as polygons. A triangle is a polygon. The figures pictured below are samples of only some types of polygons.

There are polygons with five sides, six sides, seven sides, eight sides, and more.

Polygons are named according to the number of sides. For your purposes at this time, it is enough that you know the names and recognize the more common types of polygons.

When all the sides and angles of a polygon are equal, it is called a *regular polygon*. Among those shown below, how many are regular polygons (Figure 41)?

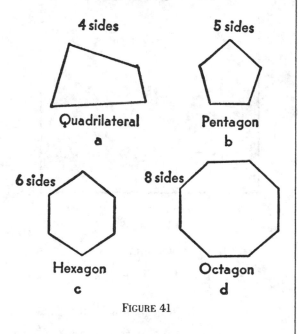

FIGURE 41

## Quadrilaterals

Among polygons with more than three sides, the quadrilaterals are the group with which you will have the most contact.

*Quadrilaterals* are polygons with four sides. There are six types of quadrilaterals as shown in Figure 42: the *rectangle*, the *square* (a special form of rectangle), the *rhomboid*, the *rhombus*, the *trapezoid* and the *trapezium*. Each of these has special features different from the others (Figure 42).

A general type among the quadrilaterals is the **parallelogram** which is any four-sided figure having the opposite sides parallel and the opposite angles equal.

A **square** is a parallelogram in which the four angles are all right angles (90°) and the four sides are all equal.

A **rectangle** is a parallelogram that has four right angles in which the opposite sides are equal.

A **rhomboid** is a parallelogram that has opposite sides parallel but no right angles.

A **rhombus** is a parallelogram that has four equal sides but no right angles.

A **trapezoid** is a quadrilateral having one pair of parallel sides.

A **trapezium** is a quadrilateral in which no two sides are parallel.

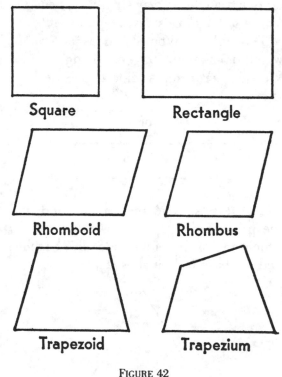

FIGURE 42

## FINDING PERIMETERS

The word "perimeter" means to measure around. It is derived from the word "peri," which means *around* and the word "meter," which means *to measure*.

### Perimeter of a Triangle

EXAMPLE 1: John's father is placing a steel fence around a triangular plot of ground which he intends to use as a garden. How many feet of fencing will he need if the dimensions are those given in the diagram below (Figure 43)?

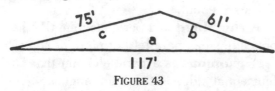

FIGURE 43

METHOD: Measure the three sides and add them together. Thus $75 + 61 + 117 = 253$ ANS.

In stating the rule for finding the perimeter of a triangle we may simply say, *add the length of the three sides*.

This rule can be stated as a *formula* which is a short way of giving a rule. We say: perimeter = side $a$ + side $b$ + side $c$

or

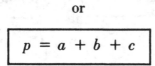

$$p = a + b + c$$

EXAMPLE 2: Give the formula and find the perimeter of this isosceles triangle in which one of the equal sides is $1\frac{1}{2}$ inches and the base is $1\frac{3}{4}$ inches (Figure 44).

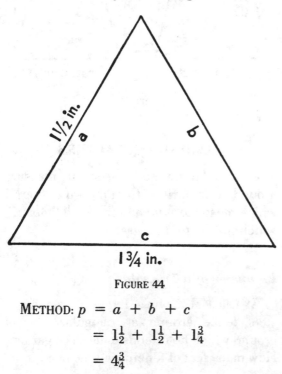

FIGURE 44

METHOD: $p = a + b + c$

$= 1\frac{1}{2} + 1\frac{1}{2} + 1\frac{3}{4}$

$= 4\frac{3}{4}$

## Perimeter of a Square

To find the perimeter of the square *ABCD* or the distance around it, add the lengths of the four equal sides or multiply one side by four (Figure 45).

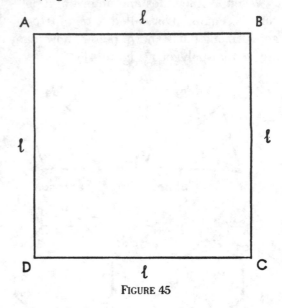

FIGURE 45

Formula for perimeter of a square is

$$p = l + l + l + l$$

or

$$p = 4l \text{ (this means } 4 \times l)$$

Measure one side of this square with your ruler and multiply by 4. What answer do you get?

EXAMPLE: What is the perimeter of this baseball diamond if the distance from home plate to 1st base is 90 feet (Figure 46)?

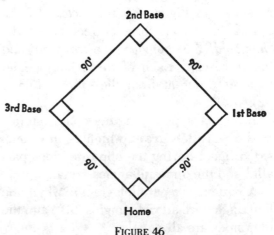

FIGURE 46

METHOD: $p = 4l$
$= 4 \times ?$
$= ?$

### Perimeter of a Rectangle or Parallelogram

In both of these figures opposite sides are equal (Figure 47).

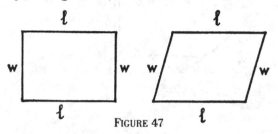

FIGURE 47

Therefore the formula for the perimeter of the rectangle or a parallelogram is:

$$p = l + w + l + w$$
or
$$p = 2l + 2w$$
or
$$p = 2(l + w)$$

NOTE: The parenthesis after the 2 in the above equation means we add $l$ to $w$, then multiply this result by the number 2. When there is no arithmetic sign between a value and a parenthesis, it is understood that multiplication is implied.

EXAMPLE: Find the perimeter of a rectangular-shaped basketball court that is 90 ft. by 50 ft. (Figure 48).

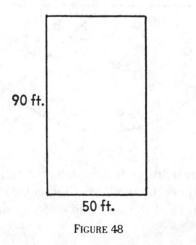

90 ft.

50 ft.

FIGURE 48

METHOD: $p = 2l + 2w$
$= 2 \times 90 + 2 \times 50$
$= 180 + 100 = 280$
or $p = 2(l + w)$
$= 2(140) = 280$

### Perimeter of a Trapezoid

In this figure (Figure 49) the 4 sides are all of different length; therefore the formula must be:

$$p = a + b + c + d$$

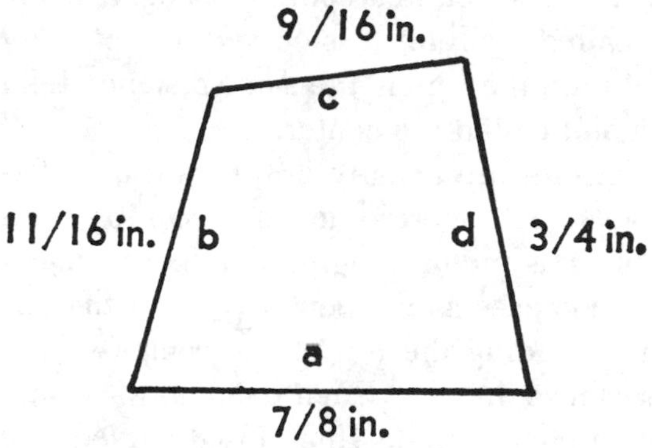

**9/16 in.**

c

**11/16 in.** b   d   **3/4 in.**

a

**7/8 in.**

FIGURE 49

$$p = \frac{7}{8} + \frac{11}{16} + \frac{9}{16} + \frac{3}{4} = 2\frac{7}{8} \text{ in.}$$

### Practice Exercise No. 104

Find the perimeters of the following figures.

**1.** A triangular plot that measures 40 feet, 50 feet and 70 feet on each of its three sides.

**2.** A square having a side of $4\frac{1}{2}$ feet.

**3.** A rectangle that is 10 feet long and $2\frac{1}{2}$ feet wide.

**4.** A triangle having sides of 135 feet, 180 feet and 225 feet.

**5.** A parallelogram that is 13 feet wide and 18 feet long.

Solve the following problems:

**6.** How many feet of chicken wire fencing will Mr. Henderson need for a rectangular yard that is $6\frac{1}{2}$ yards long and $5\frac{1}{2}$ yards wide?

**7.** In warming up, the football team is required to trot around the field twice. How far do they trot if the field is 300 ft. long and 160 ft. wide?

**8.** How many feet of baseboard is needed for a

bedroom that is $13\frac{1}{2}$ feet wide by 17 feet long in which the entrance door is 30 inches wide?

**9.** How many feet of steel rod $\frac{1}{4}$ inch in diameter is needed to make a triangular gong used for chow call on the ranch if it is an equilateral triangle 16 inches on a side?

**10.** It required 420 yards of barbed wire to fence in a rectangular-shaped cattle-grazing field. If the field is 130 yards long facing the road, what is its width?

## THE CIRCLE—A SPECIAL PLANE FIGURE

A figure which is a continuous curved line is called a circle. It is curved so that every point on the line is the same distance from a point called the center.

Circles are usually drawn with a draftsman's tool referred to as a pair of compasses as pictured here. You fix one leg of the compass as the center point of the circle. Holding the top of the compass, you twirl it so that the pencil point draws a complete curved line. This closed curved line is a **circle** (Figure 50).

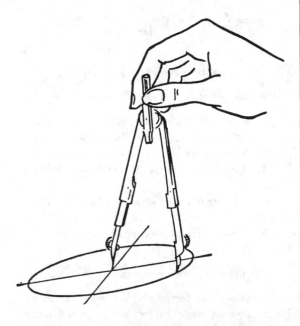

FIGURE 50

You can also draw a circle using a string tied to a pencil. You hold the free end of the string at a fixed point and rotate the

pencil, keeping the string stretched tight all the time as shown below (Figure 51).

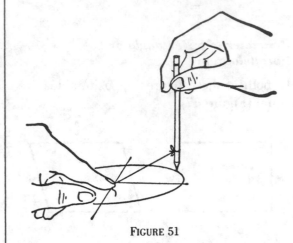

FIGURE 51

Practice drawing small and large circles with the compasses and with the string and pencil. What other material can you improvise for drawing a circle?

The continuous curved line that forms the circle is called the **circumference.**

The distance from the center point to any point on the circumference is a **radius** of that circle (Figure 52). All radii (plural) of a circle are equal.

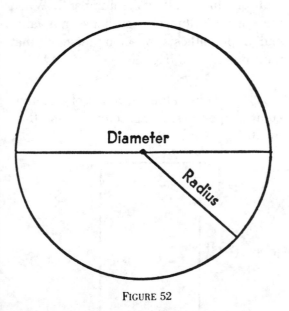

FIGURE 52

A straight line from one point on the circumference to another which goes through the center of the circle is called a **diameter.**

A diameter is twice the size of a radius.

A diameter divides a circle into two equal halves called *semicircles*.

All diameters of the same circle are equal.

| Object | Circumference | Diameter | Circumference ÷ Diameter |
|--------|---------------|----------|---------------------------|
| Clock | 16 in. | 5 in. | 3.2 |
| Pail | $32\frac{3}{4}$ | $10\frac{1}{2}$ | 3.12 |
| Tumbler | 10 in. | $3\frac{3}{16}$ | 3.13 |

### Practice Exercise No. 105

The questions which follow will test your ability to work with problems related to circles.

**1.** Draw a circle with a radius of: (a) $\frac{1}{2}$ inch, (b) $\frac{3}{4}$ inch, (c) $\frac{7}{8}$ inch; then give the size of the diameter of each.

**2.** If the diameter of a circle is $4\frac{1}{2}$ inches, what is its radius?

**3.** In the circles pictured here (Figure 53) the larger one is twice the diameter of the smaller. What is its diameter?

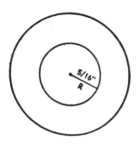

FIGURE 53

### Perimeter of a Circle

The **perimeter** of a circle is the distance around it. The circumference of any circle is its perimeter.

You can find the perimeter of any circle by measuring around it or by a formula, as we found for certain special types of polygons.

### An Experiment in Finding a Formula for Circumference

1. Measure around a can of evaporated milk. Measure the diameter across the top (as close as you can estimate).

2. Measure around a large can of fruit juice. Measure the diameter across the top.

3. Measure around a 50¢ piece. Measure across the diameter.

Make a table of your measurements as was done in the chart above.

For each set of your measurements, divide the circumference by the diameter as we have done. How much more or less than three is any one of your answers? Are they not all approximately three or a little more?

The circumference of any circle is *approximately* three times the diameter of the circle.

By measurement, the circumference of any circle is 3.1416 times its diameter.

This number, 3.1416, is thus a **constant** which represents the relationship between the circumference of any circle and its diameter.

The number 3.1416 is known as *pi* (a Greek letter) pronounced *pie* and represented by the symbol $\pi$.

Usually 3.14 is used instead of 3.1416 unless extreme accuracy is required. As a matter of fact the number of places in *pi* is much larger than four, but for our purposes we will only use two.

Here is an example showing the amount of variation obtained in using the different values for $\pi$.

EXAMPLE 1: Find the circumference of a circle that has a diameter of 14 inches.

$$
\begin{array}{cc}
\text{(a)} & \text{(b)} \\
\pi = 3.1416 & \pi = 3.14 \\
3.1416 & 3.14 \\
\underline{\times\ 14} & \underline{\times\ 14} \\
43.9824\ \text{in.} & 43.96\ \text{in.}
\end{array}
$$

(c)

$$\pi = \frac{22}{7}$$

$$\frac{22}{7} \times \overset{2}{\cancel{14}} = 44 \text{ in.}$$

**Rule: Circumference = $\pi$ × diameter.**

or

By formula: $C = \pi \times d$
$$= \pi d$$

or

$$C = \pi \times 2(\text{radius})$$
$$= 2\pi r$$

EXAMPLE 2: A piece of aluminum wrap is needed to wrap a can having a $3\frac{1}{2}$ in. diameter. What is the shortest length that can be used?

METHOD:

$$C = \pi d$$
$$= \frac{22}{7} \times 3\frac{1}{2}$$
$$= \frac{\overset{11}{\cancel{22}}}{\underset{1}{\cancel{7}}} \times \frac{\overset{1}{7}}{\underset{1}{2}} = 11 \text{ inches ANS.}$$

**Practice Exercise No. 106**

Solve the following problems.

**1.** A circular rock garden in a landscape setting is 5 feet in diameter. How many feet of protective low fencing will be needed to go around the garden?

**2.** Tape is used to bind the edge of a circular piece of metal that is 10 inches in diameter. How much tape is needed?

**3.** Joan wanted to trim the crown of a hat with ribbon. How much ribbon must she buy if the crown measures $7\frac{1}{2}$ inches in diameter?

**4.** The famous giant redwood Wawona tree in California is about 28 feet in diameter. How many feet would you walk if you went around it?

**5.** The wheel of a full size bicycle is 26 inches in diameter. How many feet does the bicycle travel when the wheel makes one full revolution? How many revolutions must it make to go a mile?

**6.** The earth has a radius of about 4000 miles. How much distance would you cover in circling the earth at the equator?

**7.** How many times would you have to go around a circular track in a mile run if the track has a diameter of 840 feet?

# CHAPTER TWELVE

# MEASURING AREAS AND VOLUMES

In finding perimeter, we measure the distance *around a plane figure*.

To measure area, we find how much *surface is taken up by a plane figure*.

Knowing just how much surface there is in a particular plane figure becomes important when we wish to cover such surfaces with paints, wallpaper, tiles, cements, carpeting, draperies, etc.

## MEASURING AREAS

Here is a problem to illustrate the difference between measurement of perimeter and surface:

John constructed two rectangular board shelves for his room. One (A) is 6 ft. long and 1 ft. wide. The other (B) is 2 ft. wide and 5 ft. long. He intends to cover them with linoleum tile squares that are 1 foot on a side. How many tiles are needed for each (Figure 54)?

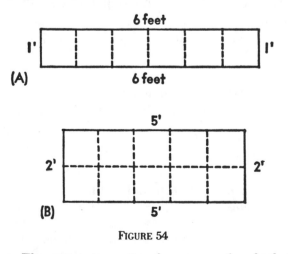

FIGURE 54

The perimeter *is the same for both* shelves (14 ft.).

To cover the shelves, A takes 6 squares and B takes 10 squares!

Since each square is 1 ft. by 1 ft. we say it has a *surface* or *area of 1 square ft.*

The *area* of A is thus 6 square feet.

The *area* of B is 10 square feet.

**Surfaces or areas are measured in square units**—*square inches, square feet, square yards.*

A *square inch* is the surface covered by a square that is 1 inch on a side.

A *square foot* is the surface covered by a square that is 1 foot on a side.

A *square yard* is the surface covered by a square that is 1 yard on a side.

Study the Table of Square Measure in Chapter 10 and commit it to memory.

To aid your understanding of **square measure,** complete the lines indicated in the square below (Figure 55). Reduced to a scale of $\frac{1}{4}$, each segment is supposed to represent 1 square inch, while the whole is 1 square foot. After you draw the cross lines, count the boxes. Do you have 144 squares?

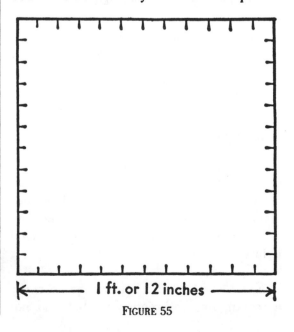

1 ft. or 12 inches

FIGURE 55

### Practice Exercise No. 107

Fill in the missing numbers.

**1.** 1 sq. ft. = _____ sq. in.

**2.** 1 sq. yd. = _____ sq. ft.

**3.** 1 sq. yd. = _____ sq. in.

**4.** _____ sq. rd. = 1 acre

**5.** 1 sq. mile = _____ acres

**6.** _____ sq. yd. = 1 sq. rod

**7.** _____ sq. ft. = 1 sq. rod

**8.** _____ sq. ft. = 1 acre

**9.** 9 sq. ft. = 1 _____ sq. yd.

**10.** 144 sq. in. = 1 _____ sq. ft.

### Finding the Area of a Rectangle

In the problem concerned with John's two shelves we found the square area by drawing in the squares and counting them. There are easier ways.

In the second board (B), how many 1 ft. squares are there along the length? *Five*.

How many 1 ft. squares are there in the width? *Two*.

The number of square feet in the total surface is 5 × 2 or 10.

We then say that *the area is 10 square feet*.

**Rule: To find the area of a rectangle,** *multiply the length by the width.*

|  | area | length | width |
|---|---|---|---|
| By formula: | $A$ = | $l$ × | $w$ |
| Area of a Rectangle | $A$ = $lw$ | | |

In computing the area of such surfaces, both dimensions must first be put in the same units. If, for example, the length is given in inches and the width in feet, one of the measures will have to be changed before multiplying.

EXAMPLE: Mrs. Rieber wanted to cover a rectangular bathroom floor but had not decided what to use for covering. She measured it with her ruler and found that it was 6 ft. long and $4\frac{1}{2}$ ft. wide. How many square feet of surface does she have to cover?

METHOD: Area of Rectangle = $lw$.
$$6 \times 4\frac{1}{2} = 27 \text{ sq. ft. area}$$

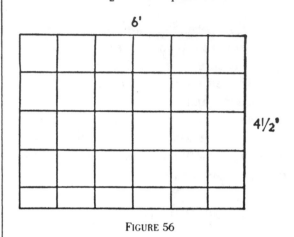

FIGURE 56

Look at the scale drawing in Figure 56. You can see that there are 6 squares across and 4 whole rows down. Then there is a row of six $\frac{1}{2}$ squares. In the 4 rows there are 6 × 4 = 24. In the last row there is 6 × $\frac{1}{2}$ = 3, 24 + 3 = 27 sq. ft.

### Finding the Area of a Square

The method for computing the area of a square is the same as for finding the area of a rectangle because a square is just a special form of a rectangle.

EXAMPLE: How much surface needs to be covered in a square ceiling that is 9 ft. on a side?

METHOD:

Length = 9 ft. and width = 9 ft.
*Therefore*     Area = 9 × 9 = 81 sq. ft.

Since both sides of a square are the same we can give the *formula* as $A = s^2$, in which the $s$ means *side* and the small number 2 written as an exponent means a number squared or multiplied by itself.

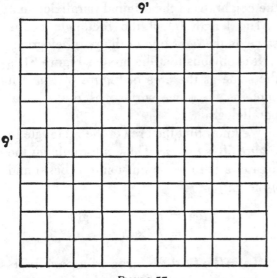

FIGURE 57

Look at the scale drawing and count the squares (Figure 57). Do you count 81?

How much is 4 squared, written $4^2$? Do you get 16? How much is $5^2$? How much is $7^2$? How much is $8^2$? Are your answers 25, 49 and 64 respectively?

### Practice Exercise No. 108

Use the formula to find the area for each of the following.

**1.** Width is 3 ft. and length is 7 ft. _____ area in sq. ft.?

**2.** Width is 1 in. and length is 2 ft. _____ area in sq. ft.?

**3.** Width is 30 ft. and length is 180 ft. _____ area in sq. yd.?

**4.** Width is 43 rd. and length is 75 rd. _____ area in acres?

**5.** Width is 6 in. and length is 12 in. _____ area in sq. ft.?

**6.** Width is 1.5 ft. and length is 8 in. _____ area in sq. ft.?

**7.** Width is 9 ft. and length is 6 ft. _____ area in sq. yd.?

**8.** Width is 11 ft. and length is 11 ft. _____ area in sq. ft.?

**9.** Width is 18 in. and length is 18 in. _____ area in sq. ft.?

**10.** Width is 16 ft. and length is 16 ft. _____ area in sq. yds.?

### Finding the Side of a Rectangle

If you know the area of a rectangle and one side, you can find the other by division.

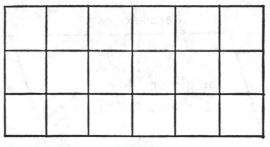

FIGURE 58

EXAMPLE 1: The rectangle shown here (Figure 58) represents 18 square inches. If it is 6 inches long, how wide is it?

If you counted the squares, you would find the width to be 3 inches. As a check $l \times w = A$; $3 \times 6 = 18$.

Without counting the squares, you can find the width by dividing 18 by 6 because if $3 \times 6 = 18$ then $18 \div 6 = 3$. We can now state this as a rule.

**Rule: To find the length or width of a rectangle when the area and one dimension are given,** *divide the Area by the given dimension.*

By formula: If $A = l \times w$

Then $w = A \div l$ or $w = \dfrac{A}{l}$

This is stated as *width is equal to the area divided by the length.*

And $l = A \div w$ or $l = \dfrac{A}{w}$

This is stated as *length is equal to the area divided by the width.*

EXAMPLE 2: If the area of a rectangular field is 2400 square feet and it is 30 feet wide, how long is it?

METHOD: $l = A \div w$

$l = 2400 \div 30$ or $\dfrac{2400}{30}$

$= 80$ ft. ANS.

### Finding the Area of a Parallelogram

Copy, on a separate sheet of paper, the parallelogram drawn below (Figure 59).

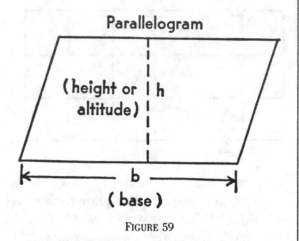

FIGURE 59

From the corner *A*, draw the line *AE* so that it will make a right angle with the opposite side. This line is called the height (*h*) or altitude of the figure.

Cut off the right triangle *AED* (Figure 60), formed by the line *AE* and place it on the opposite side to form a *rectangle*.

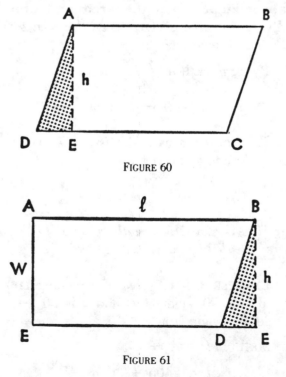

FIGURE 60

FIGURE 61

Notice that the width (*w*) of the newly formed rectangle (Figure 61) is the same as

the height (*h*) of the original parallelogram.

The length (*l*) of the rectangle is the same as the *base* line of the parallelogram.

It is obvious that the area of Figure 61 is the same as the area of Figure 60 since it contains the same surface as the original parallelogram.

We know that the area of the rectangle in Figure 58 is $l \times w$. Thus, a formula for the area of a parallelogram, using *b* (base) and *h* (height) is

$$\text{Area} = \text{base} \times \text{height}$$
$$A = b \times h$$

**Rule: To find the area of a parallelogram,** *multiply the base by the height*.

EXAMPLE: What is the area of a parallelogram that has a base of 5 feet and a height of 6 feet (Figure 62)?

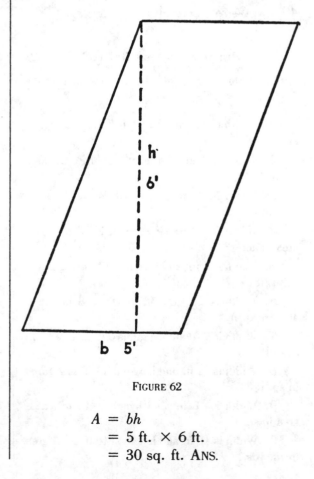

FIGURE 62

$$A = bh$$
$$= 5 \text{ ft.} \times 6 \text{ ft.}$$
$$= 30 \text{ sq. ft. ANS.}$$

**Practice Exercise No. 109**

Given the base and height of the parallelograms below, find the areas as directed.

| | Base | Height | Area |
|---|---|---|---|
| **1.** | 3 ft. | 1 yd. | ? sq. ft. |
| **2.** | 25 in. | 18 in. | ? sq. ft. |
| **3.** | 5 ft. | 9 ft. | ? sq. ft. |
| **4.** | 18 ft. | 4 yd. | ? sq. yd. |
| **5.** | 18 ft. | 4 yd. | ? sq. ft. |

## Finding the Area of a Triangle

Study the illustration below in Figure 63. You can see that if we duplicate the right triangle as shown, we form a rectangle. This shows that the area of the triangle is *half* the area of the rectangle. Now look at Figures 64 and 65; here we see that the triangle is *half* the area of the parallelogram.

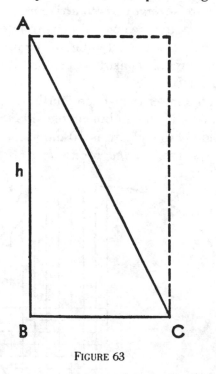

FIGURE 63

You can prove this to yourself. Draw a triangle of any shape. Cut out two copies of it. You can place them together in such a way as to form either a rectangle or a parallelogram.

If the triangle is a right triangle, you obtain a rectangle as in Figure 63.

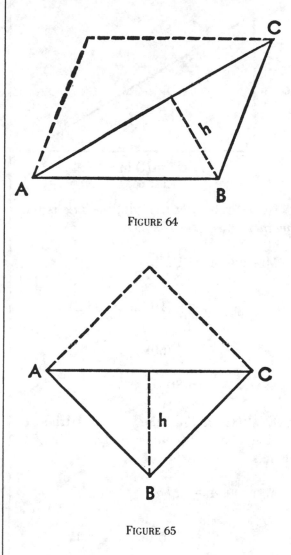

FIGURE 64

FIGURE 65

If the triangle is obtuse or acute, you obtain a parallelogram as in Figures 64 and 65.

Notice the *height* lines drawn in Figures 64 and 65. This is *the perpendicular distance from a vertex to the opposite base*. No height line is needed in Figure 63. Why?

From the facts shown above, you can readily see that if the area of a parallelogram equals base × height, as previously shown,

*The area of a triangle equals*
$$\frac{1}{2} \text{ base} \times \text{height or } A = \frac{1}{2}bh.$$

EXAMPLE 1: Find the area of the *right* triangle shown here (Figure 66).

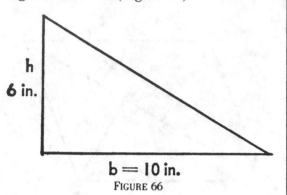

FIGURE 66

NOTE: In a *right* triangle, the *side* is the *height*.

METHOD: $A = \frac{1}{2} bh$.

$$= \frac{1}{2} (10 \text{ in.} \times 6 \text{ in.})$$

$$= \frac{1}{2} (60)$$

$$= 30 \text{ sq. in.}$$

EXAMPLE 2: Find the area of a triangle having a base of 6 inches and a height of 4 inches.

METHOD: $A = \frac{1}{2} bh$

$$= \frac{1}{2} (6 \text{ in.} \times 4 \text{ in.})$$

$$= \frac{1}{2} (24)$$

$$= 12 \text{ sq. in.}$$

**Practice Exercise No. 110**

Find the areas of the following triangles.

**1.**

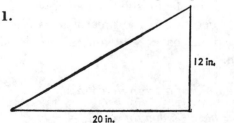

**2.**

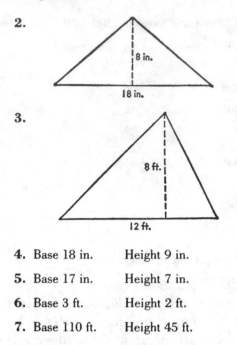

**3.**

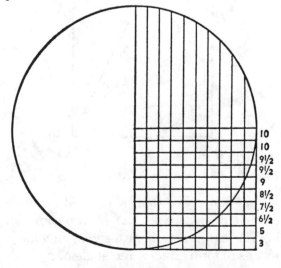

**4.** Base 18 in.      Height 9 in.

**5.** Base 17 in.      Height 7 in.

**6.** Base 3 ft.      Height 2 ft.

**7.** Base 110 ft.      Height 45 ft.

**8.** Base 100 ft.      Height 50 ft.

### Finding the Area of a Circle

There are several ways to arrive at the rule for finding the area of a circle. An interesting method is illustrated below (Figure 67).

This circle is constructed so that it has a radius of 10 units.

On the lower right-hand fourth, a square is drawn with the radius as the side. Since the side of the square is 10 units, *the area of the square on the radius is $r^2$ or 100 square units.*

FIGURE 67

Count the units that are within the marked fourth of the circle. You will have to count fractions of units in the square near the border of the circle. We estimate the number of squares in each row as follows: 10, 10, $9\frac{1}{2}$, $9\frac{1}{2}$, 9, $8\frac{1}{2}$, $7\frac{1}{2}$, $6\frac{1}{2}$, 5, 3— this gives $78\frac{1}{2}$ square units in one-fourth of the circle. In the whole circle you should have $78\frac{1}{2} \times 4$ or approximately 314 square units.

In comparing this approximate area of the circle (314 square units) with the area of the square on the radius (100 square units) we get $314 \div 100 = 3.14$. This, you will recall is the same ratio ($\pi$) found in connection with the circumference of a circle. This leads to the rule:

*The area of a circle is 3.14 or $\pi$ times the square of the radius.*

By formula:

$$A = 3.14 \times r^2 \text{ or } \frac{22}{7} \times r^2$$

Using $\pi$ for 3.14 the formula for area of a circle is:

$$A = \pi r^2$$

EXAMPLE 1: Find the area of a circle whose radius is 6 inches.

METHOD:
$$A = \pi r^2$$
$$A = 3.14 \times (6 \times 6)$$
$$A = 3.14 \times 36 = 113.04 \text{ inches}$$

EXAMPLE 2: Find the number of square inches of copper in a circular serving tray that has a diameter of 14 inches.

METHOD: $A = \pi r^2$

$$A = \frac{22}{7} \times (7 \times 7)$$

$$= \frac{22}{7} \times \overset{7}{\cancel{49}} = 154 \text{ sq. in.}$$

*Step 1*. $r = \frac{1}{2}$ diameter or $\frac{14}{2} = 7$.

*Step 2*. Substitute in the formula.

*Step 3*. Multiply $\pi$ by the radius squared.

### Practice Exercise No. 111

Find the area of these circles, using $\frac{22}{7}$ for $\pi$.

1. Radius is 4.9 ft.

2. Diameter is 5.6 in.

3. Radius is 20 in.

4. Diameter is 42 in.

5. Radius is 35 ft.

6. Radius is 18 in.

7. Diameter is 28 in.

8. Radius is 21 ft.

### Practice Exercise No. 112

The problems below will test your ability to work with plane figures.

1. Mr. Wright purchased a square lot of 160 ft. on a side. He divided it into plots of 5 ft. × 8 ft. for bath houses. How many plots were there?

2. Lyle's father bought a television set with a circular tube 24 inches in diameter. How many square inches is the surface of the tube?

3. Mrs. Harmony plans to use 9" × 9" linoleum tiles to cover the kitchen floor. The room is 18 ft. long and 12 ft. wide. How many tiles will she need?

4. Weldon decided to make triangular banners announcing a sale. They were to have a base of 12 inches and to be 18 inches high. How many square feet of material does he need for 10 banners?

5. Terry's father bought a rectangluar piece of land in the country. It measured 120 ft. by 400 ft. How much more or less than one acre did he have?

6. Alfred decided to go into the business of making triangular corner shelves in mass production. He decided they should be 8 inches in height with a base of 12 inches. How many can he make from 200 sq. ft. of lumber?

7. The sail on Paul's boat tore to shreds in a wind. It has a height of 15 ft. and a base of 9 ft. How many square feet of sail material will he need to replace the sail?

**8.** Jill's father built a circular pool 15 ft. in diameter and a walk 3 ft. wide around it. What is the area of each?

**9.** For an election campaign, James made a circular poster with a diameter of 3 inches. Ben, his opponent, made his with a diameter of 6 inches. What is the area of each, and how much larger is Ben's than James'?

**10.** A rug cleaning company charges 15¢ a square foot to clean rugs in the home. How much will it cost Mrs. Brown to have her circular-shaped rug, which measures 10 ft. in diameter, cleaned?

## MEASURING VOLUME

How much does it hold? How much space does it occupy?

When you ask these questions you are referring to the *volume* or *capacity* of an object.

When we spoke of surface or area, we referred to plane figures because they took up only *one plane* or *flat surface*.

When we speak of **volume,** we refer to *solid* figures. These are figures with more than one plane. While plane figures have only two dimensions, solid figures have the added dimension of depth.

Think of volume as referring to containers of every shape and size.

A huge warehouse is a container. What may it contain?

A coal bin is a container.

An oil drum is a container.

Every room in your house is a container.

A grocery carton is a container.

A can of soup is a container.

An ice cream cone is a container.

The measuring cups of Chapter 1 are containers.

When you determine how much the container will hold, whether it is oil in the drum, ice cream in the cone, or soup in the can, you are measuring its *volume*, or **capacity** as we call it.

Volume is expressed in *cubic units*, such as the cubic inch, cubic foot, or cubic yard.

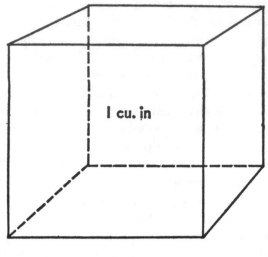

**I cu. in**

FIGURE 68

A **cube** is a solid shape with six square sides in which the length, width and height are equal and all of the angles are right angles. It may also be called a **rectangular solid** or **prism** (Figure 68).

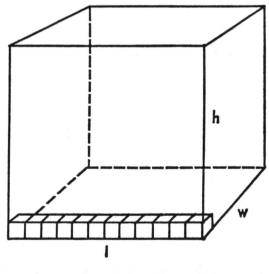

**h**

**w**

**l**

FIGURE 69

A **cubic inch** is the space occupied by a cube 1 inch long, 1 inch wide, and 1 inch high.

A **cubic foot** is the space occupied by a cube 1 foot long, 1 foot wide, and 1 foot high.

How many cubic inches would there be in a cubic foot? One way we can determine this is to take a figure equal to 1 cubic foot and place in it cubes that are 1 cubic inch in size.

Look at the illustrations. You can see that 12 cubes would make 1 row across the length of the bottom (Figure 69).

It would take 12 × 12 or 144 cubes to make one layer in length and width across the bottom (Figure 70).

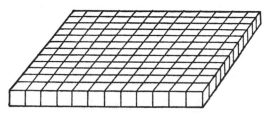

FIGURE 70

To fill the one foot cube completely it would take 12 layers or 12 × 144 or 1728 cubic inches. By looking at Figure 71 you can see that you can count 1728 cubic inches in 1 cubic foot.

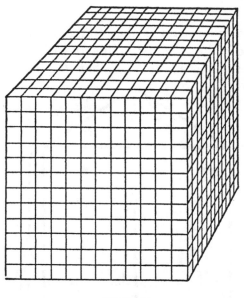

FIGURE 71

Naturally, it is not practical to count cubic units in order to find the cubic capacity of any object. There is an easier way to do this by arithmetic. Actually, we did it in this example. Check back and you will find the rule.

**Rule: To find the volume of a cube,** *multiply the length by the width by the height.*

By formula:

$$V = lwh$$

This rule or formula applies to any solid that is *square or rectangular* such as the ordinary gift box, carton or rectangular shaped room.

EXAMPLE: Applying the rule, find the volume in cubic inches of the rectangular box drawn to scale which is 6 inches long, 3 inches wide and 4 inches high (Figure 72).

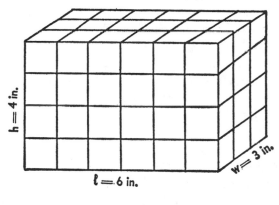

FIGURE 72

METHOD: $V = lwh$
$V = 6 × 3 × 4$
The volume = 72 cu. inches

Check: By counting the boxes in the illustration.

Note that in finding volumes, it is necessary to get the dimensions of any object into similar units before multiplying. Generally, it is easier to change the dimensions to correspond with the units desired for the answer. If, for example, you measure a concrete pavement in feet, but intend to buy ready-mix cement to do the paving, you will want your dimensions in yards because ready-mix cement is sold *by the cubic yard*.

### Practice Exercise No. 113

Problems in finding volumes of rectangluar solids.

**1.** A carton is 18 inches long, 12 inches wide, and 8 inches deep. What is its capacity in cubic inches?

**2.** A truck body is 10 ft. long, $6\frac{1}{2}$ ft. wide, and 4 ft. deep. What is its capacity?

**3.** Mr. Lyle bought a sandbox for his children which measured 3 ft. by 3 ft. by 8 inches deep. How many cubic feet of sand must he order to fill it?

**4.** The closet in Weldon's room is 3 ft. by 3 ft. and 8 ft. high. How many pounds of moth flakes will be needed if one pound protects 36 cubic feet?

**5.** What is the capacity in cubic feet of a freezer that is 32 inches wide, 36 inches long, and 27 inches high?

**6.** A contractor charges $20 a cubic yard for labor and materials to make a driveway. How much will Mr. Quigg have to pay for a driveway 8 ft. wide, 50 ft. long and 6 inches thick?

**7.** A candy manufacturer makes fudge squares 1 inch by 1 inch by $\frac{1}{2}$ inch thick. How many pieces can he pack in a box that is 6 inches long, 4 inches wide, and 2 inches deep?

### Another Way of Figuring the Volume of a Rectangular Solid

We know from the measurement of surfaces that the area of the *base* of a rectangular shaped figure is found by multiplying length times width. Thus $A = lw$.

Since the *volume* of a rectangular prism is found by multiplying length × width × height, we could write the formula as: $V = A \times h$. $A$ stands for *area of the base* and $h$ for the *perpendicular height of the prism*.

From this we derive an important and helpful rule about finding the volume of any prism.

Before stating the rule, we should define the word prism. By definition: A **prism** *is a solid figure whose top and bottom bases are equal polygons and whose other faces are parallelograms*.

**Rule: The volume of a prism is equal to the area of the base × the height.**

$$V = Ah$$

EXAMPLE: Find the volume of the illustrated rectangular prism if the area of the base is 40 sq. inches and the height is 3 in. (Figure 73).

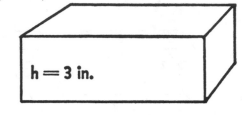

**h = 3 in.**

**base = 40 sq. in**

FIGURE 73

METHOD:

$V = Ah$
$\phantom{V} = 40 \times 3 = 120$ cu. inches

### Finding the Volume of a Triangular Prism

Using the formula for the volume of a rectangular solid in the form given above, we can apply it in finding the volume of other prisms as illustrated here (Figure 74).

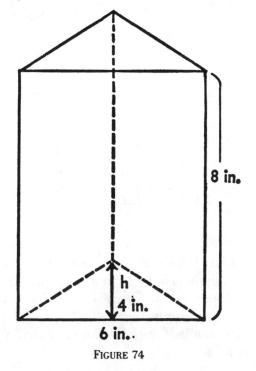

8 in.

h

4 in.

6 in.

FIGURE 74

EXAMPLE 1: Find the volume of the triangular prism illustrated in Figure 74. The base line of the lower triangle is 6 inches and the altitude of the lower triangle $h$ is 4 inches, while the perpendicular height of the prism is 8 inches.

METHOD: Find the area of the base triangle.

$$A = \frac{1}{2}bh$$

$$= \frac{1}{2}(6 \times 4) = 12 \text{ sq. in.}$$

The volume of the prism is the area of the base times the height.

$$V = Ah$$

$$= 12 \times 8 = 96 \text{ cu. in.}$$

*Step 1.* Find the area of the triangular base.

*Step 2.* To find the volume, multiply the area of the base by the perpendicular height of the prism.

EXAMPLE 2: Find the volume of a triangular prism having a base with dimensions of $b = 8$ inches, $h = 4\frac{1}{2}$ inches, and the height of the prism is 10 inches.

METHOD:

$$V = Ah \text{ in which } A = \frac{1}{2}bh$$

$$A = \frac{1}{2} \times 8 \times \frac{9}{2} = \frac{72}{4} = 18 \text{ sq. in.}$$

$$V = 18 \times 10 = 180 \text{ cu. in.}$$

### Finding the Volume of a Cylinder

It is not too difficult to demonstrate how the same formula used for finding the volume of rectangular and triangular prisms can be applied in finding the volume of a cylinder.

Volume = Area of base × height.

Since the base of the cylinder is a circle,

$$A = \pi r^2$$

Formula for volume of a cylinder is therefore:

$$V = \pi r^2 h$$

EXAMPLE: Find the volume of the cylinder pictured in Figure 75 if it has a radius of $1\frac{1}{2}$ inches and a height of 28 inches.

METHOD: $V = \pi r^2 h$

$$\pi = \frac{22}{7}, r = \frac{3}{2} \text{ in., } h = 28 \text{ in.}$$

$$V = \frac{22}{7} \times \left(\frac{3}{2}\right)^2 \times 28$$

$$= \frac{22}{7} \times \frac{9}{4} \times 28 = 198 \text{ cu. in.}$$

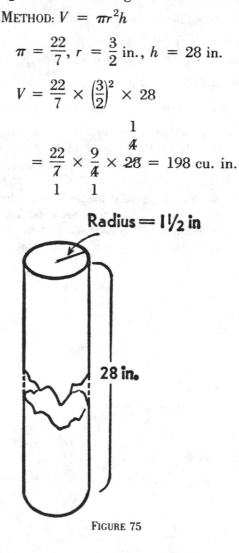

FIGURE 75

**Practice Exercise No. 114**

Find the volume of the given figures.

| Figure | Base Information | Height |
|---|---|---|
| 1. Cube | Area = 16 sq. in. | 4 in. |
| 2. Triangular prism | Area = 9 sq. in. | 7 in. |
| 3. Rectangular prism | Area = 30 sq. in. | 1 ft. |
| 4. Hexagonal prism | Area = 120 sq. in. | 4 in. |

**5.** Cylinder       Area = 48 sq. in.        8 in.

**6.** Cylinder       Diameter 4 in.           6 in.

**7.** Rectangular    8 in. long, 7 in. wide   9 in.
       prism

**8.** Triangular     Base 6 in., alt. 6 in.   8 in.
       prism

**9.** Cube           2 ft. long, 2 ft. wide   2 ft.

**10.** Cylinder      7 ft. radius             18 in.

Problems: Give answers to nearest whole number.

**11.** The local utility company erected a cylindrical tank to store cooking gas. If it has a diameter of 200 feet and stands 91 feet high, what is the capacity of the tank?

**12.** Gasoline is stored at the refinery in cylindrical tanks that have a base diameter of 60 feet and stand 40 feet high. If $7\frac{1}{2}$ gallons take up 1 cubic foot, how many gallons can be stored in each tank?

**13.** A stainless steel cylindrical tank is used to transport milk. The tank is 20 ft. long and has an inside diameter of 5 ft. How many cubic feet of milk does it carry? If milk weighs about 67 lb. per cubic foot, how much does the full load weigh?

**14.** On a cattle ranch they constructed a water trough in the form of a triangular prism that extended for 90 ft. The base was a right triangle with a base of 4 ft. and the perpendicular side 5 ft. Figure a cubic ft. of water as $7\frac{1}{2}$ gal. How many gallons will the trough hold? (Figure 76).

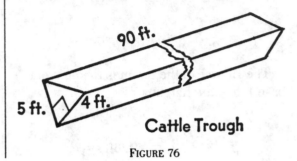

**Cattle Trough**

FIGURE 76

# CHAPTER THIRTEEN

# COMPARISON OF QUANTITIES BY RATIO AND PROPORTION

## RATIO

Pete's little league team won 15 games and lost 5 games.

We can compare these two number facts in several ways.

(a) Pete's team won *10 games more* than they lost.

(b) They won *three times as many* games as they lost.

(c) Pete's team won three fourths of the games it played.

In (a) we compared the two numbers by subtraction: $15 - 5 = 10$.

In (b) and (c) we compared the two numbers by the process of division.

(b) $15 \div 5 = \dfrac{3}{1}$    (c) $15 \div 20 = \dfrac{3}{4}$

When two numbers are compared by the process of division, the result is called the *ratio* of one quantity to the other.

**Ratio** means relationship.

### How to Indicate Ratio

(a) A ratio may be indicated as a fraction. The ratio between the games won and the total games played is $\frac{15}{20}$. It can be stated as *15 to 20* or *3 to 4*. These are equivalent forms.

(b) The same ratio may be written with a colon between the quantities compared— $15 : 20$ or $3 : 4$. It is still stated as 3 to 4. The colon (:) is an abbreviation of the division sign ($\div$).

When writing a ratio, you must be careful about which number is written first. *Write the number asked about first*, then the number with which it is being compared.

This can be clearly demonstrated by the performance of Pete's little league team. If the question is how many games did they win out of those played, you first write games won and then games played. Thus $15 : 20$ or 3 out of 4.

If the question is how many games did they lose out of those played, you first write games lost and then games played. Thus $5 : 20$ or 1 out of 4.

The quantities compared must be expressed in the *same units* as in ordinary division. We compare feet with feet and not with inches.

Reduce fractions to lowest terms.

Ratios are written without units or dimensions.

As is true for fractions, *both terms (numbers) of a ratio may be multiplied or divided by the same number without changing the value of the ratio*.

### Practice Exercise No. 115

In the following exercise write the ratio of the first number to the second number as a fraction. Write it first in the higher form and then reduce the ratio to lowest terms.

1. 2 inches to 1 ft.

2. 3 minutes to 1 hour

3. 2 quarts to 10 pints

4. $1.00 to $5.00

5. 3 pints to a gallon

6. 15 to 5

7. 24 to 12

8. 72 to 108

9. 5 to $\frac{1}{2}$

10. $8\frac{1}{2}$ to $\frac{1}{2}$

Using the colon, express the items below as ratios. First write them in higher form and then reduce them to lowest terms.

**11.** 5 in. to 10 in.

**12.** 2 oz. to 2 lb.

**13.** 20¢ to 80¢

**14.** $4.00 to 50¢

**15.** 10 sec. to 1 min.

**16.** 2 days to 12 hours

**17.** 3 oz. to 1 lb. 5 oz.

**18.** $\frac{5}{6}$ min. to 20 sec.

**19.** 45 percent to 90 percent

**20.** .2 inch to .8 inch.

### Practice Exercise No. 116

Solve the following problems pertaining to ratios.

**1.** Henry is 12 years old and his mother is 34. What is the ratio of his age to his mother's?

**2.** Michael saved $10 of the $40 that he earned. What is the ratio of the money he saved to the money he earned? What is the ratio of the money saved to the money spent?

**3.** On a vocabulary test of 100 words, Alfred received a mark of 80%. What is the ratio of the number he answered correctly to the total number?

**4.** A chemical for killing weeds contained directions which indicated a mixture of 4 oz. to a pint of water. What is the ratio of the chemical to the total amount of liquid to be used for spraying?

**5.** On a map, one inch is indicated as representing 10 miles. What is the ratio of the distance on the map to the actual distance?

## PROPORTION

A **proportion** is a method of expressing *equality between two ratios*.

The equation between the two ratios may be indicated by the double colon or proportion sign (::) or with the sign of equality (=).

For example, 1 : 3 :: 2 : 6 is a proportion that is read, *1 is to 3 as 2 is to 6*. It may also be written as $\frac{1}{3} = \frac{2}{6}$.

In any proportion, as in this one: 1 : 3 :: 2 : 6, the first and last terms (1 and

6) are called the **extremes** while the second and third terms (3 and 2) are called the **means**.

Multiply the extremes $1 \times 6 = $ ?

Multiply the means $3 \times 2 = $ ?

Compare the products. Are they equal? Will they always be equal? Yes, if it is a true proportion.

**Rule: In a proportion,** *the product of the means is equal to the product of the extremes*.

If you write the proportion as equal fractions, in this form $\frac{1}{3} \diagup\!\!\!\!\diagdown \frac{2}{6}$, note that the means and extremes are diagonally opposite each other and it follows that *cross products of a proportion are always equal*.

If you can make a proportion out of a problem, the rule offers you an easy solution by the following method which is a procedure used in elementary algebra.

EXAMPLE 1: If three postcards cost 10 cents, how much would 12 cards cost?

METHOD: The ratio of 3 cards to their cost, 10¢, should be proportionate to the ratio of 12 cards to their cost.

Thus, 3 : 10 :: 12 : ? We have to find the value of the missing term. The letter $x$ is traditionally used to denote a missing term or an unknown quantity. Rewriting the proportion we get:

$$3 : 10 :: 12 : x$$

1. 3 times $x$ = 10 times 12
   or $3x = 120$

2. $\dfrac{3x}{3} = \dfrac{120}{3}$

$x = 40$ ANS.

*Step 1.* Product of the extremes equals the product of the means.

*Step 2.* Divide both sides by the number that is the multiplier of $x$. Both sides of an equation may be divided by the same number without changing the value of the equation.

EXAMPLE 2: A picture that is 6 inches long by $2\frac{2}{3}$ inches wide is to be enlarged so that it will be 9 inches long. How wide must it be to maintain the same proportions as the original?

METHOD: Ratio of original length to width is $6 : 2\frac{2}{3}$. The proportionate ratio of new length to width is $9 : x$.

Thus $6 : 2\frac{2}{3} :: 9 : x$

1. 6 times $x = 2\frac{2}{3}$ times 9
   or $6x = 24$

2. $\dfrac{6x}{6} = \dfrac{24}{6}$

   $x = 4$ ANS.

*Step 1.* Product of extremes equals product of means.

*Step 2.* Divide both sides by the number that is the multiplier of $x$.

The above process is the **equation** method of solving problems containing an unknown. As we noted, it is a method which employs elementary algebra. Although this is the preferred method, there is also an arithmetic method for solving proportions.

To use a strict arithmetic procedure for finding the missing term in a proportion, you may employ the following rule.

**Rule: In a proportion,** *the product of the means divided by either extreme, gives the other extreme as the quotient.* The converse is also true.

TO ILLUSTRATE: $2 : 5 :: 6 : 15$.
$$5 \times 6 = 30, \; 30 \div 15 = 2 \text{ or}$$
$$30 \div 2 = 15$$

Thus in Example 2 above, $6 : 2\frac{2}{3} :: 9 : ?$ Multiply the means $2\frac{2}{3} \times 9 = 24$; divide the product by the known extreme; $24 \div 6 = 4$. The quotient is the unknown term.

**Practice Exercise No. 117**

Find the missing term.

1. $2 : 4 :: 4 : \underline{\hspace{1cm}}$
2. $3 : 11 :: \underline{\hspace{1cm}} : 22$
3. $8 : \underline{\hspace{1cm}} :: 3 : 9$
4. $7 : 3\frac{1}{2} :: 14 : \underline{\hspace{1cm}}$
5. $\underline{\hspace{1cm}} : 4\frac{1}{2} :: 6 : 12$

Use the proportion formula to solve the problems below.

6. A 6 oz. can of frozen orange juice sells for 26¢. What should be the price of an 18 oz. can at the same rate?

7. An antifreeze solution for an automobile is to be used in the ratio of 1 part of antifreeze to 2 parts of water. If 2 gallons of antifreeze are required, how much water is needed?

8. In Roland's 4-H Club there are 36 girls. The ratio of boys to girls is 2 to 3. How many boys are there?

9. Mr. Lamb's car will run 80 miles on 5 gallons of gasoline. How many gallons will he need in order to travel 200 miles?

10. If four newspaper boys working at capacity can deliver 320 newspapers in one afternoon, how many boys, working at the same pace will be needed to deliver 400 papers?

# CHAPTER FOURTEEN

# GRAPHS—PICTURES OF NUMBER COMPARISONS

It is said that one picture is worth a thousand words. Graphs are pictures.

They are used to illustrate vividly and graphically all kinds of number facts and comparisons. The word graph is a shortened form of the word graphic which means to illustrate by pictures.

In education and business the use of many different types of graphs is standard practice. The most common types are: bar graphs, line graphs, pictographs, rectangle graphs, and circle graphs.

## BAR GRAPHS

A popular and easy type of graph to read and make is the **bar graph** (Figure 77). Generally, it is used to compare quantities and to illustrate growth, improvement or a trend. The significant quantities are represented by the length or height of a bar. For this reason it is called a bar graph.

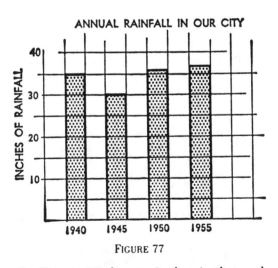

FIGURE 77

In Figure 77 the *vertical axis* shows the number of *inches of rain*. The *horizontal axis* shows the years at *five-year intervals*.

If the bars of the graph are vertical, it is referred to as a vertical graph, and if the bars are horizontal, it is called a horizontal graph.

In Figure 77 you see an example of a *vertical bar graph* showing a comparison of the amount of average annual rainfall in one city at five-year intervals. Notice that the graph has a title. Look for graphs in magazines. You will find that almost always they contain a descriptive title.

The same graph can be constructed as a *horizontal bar graph* with the comparison of the amounts of rainfall on the horizontal scale and the five-year intervals on the vertical axis as shown in Figure 78.

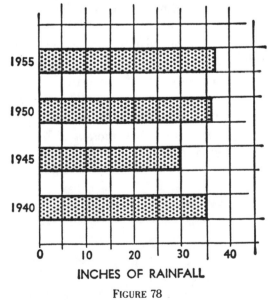

FIGURE 78

*Interpretation*

What information can you obtain practically at a glance from such a graph?

1. Which was the driest year?
2. What was the amount of rainfall in the driest year?
3. Which year had the most rainfall?
4. How much difference was there between the driest and the wettest year?
5. What would be a reasonable estimate of the expected rainfall in any one year?

If you read in the newspaper that the United States Government spent 820 billion dollars in 1984 while it only spent 500 and 700 billion dollars for the years 1980 and 1982, respectively, you know that there is an increase, but you don't see the trend.

In Figure 79 is a *bar graph* taken from United States Governmental Statistics. From this graph it becomes much clearer how expenditures are rising from year to year. Just looking at the numbers the rate of increase is not as apparent as it is with the graph.

### U.S. GOVERNMENTAL EXPENDITURES IN BILLIONS OF DOLLARS

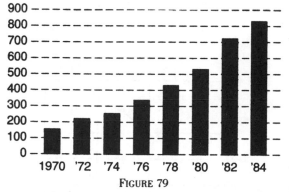

FIGURE 79

What do you learn from the graph in Figure 79?

Is the amount of money spent increasing the same each year?

Do you expect the amount of increase from 1984 to 1986 to be only 100 billion dollars, as it was between 1982 and 1984?

In Figure 80 we see a *bar graph* that has comparative data on the same graph.

Popular Vote Cast for President, by Major Party: 1952 to 1980

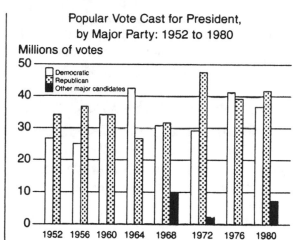

FIGURE 80

Who won the election in the year 1972, Republican or Democrat?

Are more people voting in later years of the graph?

Does it look like the Independent votes were taken from the Democrats or the Republicans?

Who won the election in 1960?

### Practice Exercise No. 117B

An automobile manufacturer had this graph in the Instruction Booklet for new car owners (Figure 81).

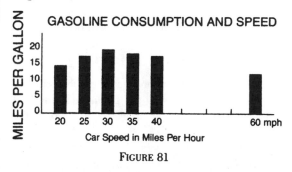

FIGURE 81

Answer these questions based on the graph.

**1.** At what speed was the lowest gas mileage obtained?

**2.** At what speed was the best gas mileage obtained?

**3.** What was the gas mileage at 60 mph?

**4.** How many miles more per gallon can you get at a speed of 30 mph than at a speed of 20 mph?

**5.** If we assumed that the gasoline consumption *increased* at a steady rate as the speed increased from 30 mph. to 60 mph., what would be the miles per gallon at a speed of 50 mph.?

### How to Construct the Bar Graph

Decide first whether the bars are to be vertical or horizontal. This will depend upon the available space and possibly eye appeal.

Select a scale with intervals so that the largest bar will almost fill it.

Start the scale at zero.

The bars should be of equal width and the spaces between them should be of equal width.

#### Suggestions for Home Study Practice

**1.** At a summer camp, the campers were asked to vote for their first choice of a mid-morning athletic activity, with the following results: swimming 75 votes, baseball 55 votes, basketball 50 votes, tennis 25 votes, water skiing 40 votes. Construct a vertical bar graph showing the results of the voting. Make up a suitable title for the graph.

**2.** Make a horizontal bar graph to illustrate the following information. Round off to the nearest thousand. Add a title.

| Ocean | Ocean Depth |
|---|---|
| Arctic Ocean | 3,953 ft. |
| Atlantic Ocean | 12,880 ft. |
| Indian Ocean | 13,002 ft. |
| Pacific Ocean | 14,048 ft. |

**3.** Construct a horizontal bar graph, with an appropriate title, showing the population growth of your state at 10-year intervals from 1900 to the present.

## LINE GRAPHS

The line graph takes its name from the fact that either straight connecting lines or curved lines are used to show the number relationships.

The line graph is most applicable in showing how relationships *change*. It is most helpful when we want to illustrate increasing or decreasing quantities.

### Characteristics of the Line Graph

The line graph has two scales.

The scales do not have to start at zero.

The value of the spaces or intervals of each scale is selected to fit the size graph desired.

The graph is a series of dots or points that are connected by lines. It is made as if the tops of a vertical bar graph were joined by lines and the bars erased.

Look at the line graph in Figure 82. Notice how the line goes up *(slope)*. The slope of the line gives an indication of how fast things are changing. It conveys this information much better than a bar graph.

### How to Construct a Typical Line Graph

If we wanted to show the progress of the average weight increase of boys from age 8 to 14, we would use the type of line graph shown in Figure 82. The facts are given in the table below this graph.

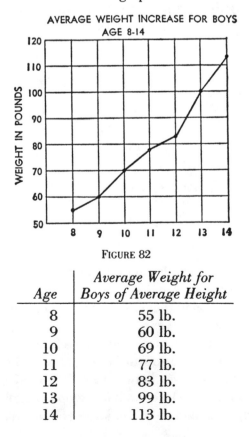

FIGURE 82

| Age | Average Weight for Boys of Average Height |
|---|---|
| 8 | 55 lb. |
| 9 | 60 lb. |
| 10 | 69 lb. |
| 11 | 77 lb. |
| 12 | 83 lb. |
| 13 | 99 lb. |
| 14 | 113 lb. |

### Steps to Follow in Constructing a Line Graph

1. Decide which scale is to be the vertical axis. Since the age progresses at a uniform rate, and *weight* is the varying item, you would make weight the scale on the left or the vertical axis. The age factor will then be on the horizontal scale.

2. Decide the intervals or size of the steps needed in each scale. To do this, take the highest number on the scale and subtract it from the lowest. For the *age scale*, you have 14 − 8 or 6. Since there will be a range of only size years between the lowest and highest age we let each year represent one space or step.

Using the same procedure for the weight scale, you have 113 − 55 or a difference of 58 pounds between the lowest and greatest weight. Here you could not let each space equal one pound because there would not be enough room on the page to make the scale. Therefore, you have to decide on a convenient interval. In this case 10 pounds per interval seems logical, because it would require about seven spaces. This is so because we always start a little below the lowest number and go a little above the highest. Thus the weight scale reads from 50 pounds to 120 pounds at 10-pound intervals.

3. Next you proceed to locate the points on the graph after drawing in the vertical and horizontal lines.

At the line for each age level, you go up the vertical scale and place a dot on the *age line* that represents the corresponding weight on the vertical scale. Thus for age 8 the corresponding weight is 55. This is exactly midway between 50 and 60 on the vertical scale. The first dot is placed at this point. Next move over to the line for age 9. The corresponding weight (from the table) is 60. This falls exactly on the 60 line for weight, and the second dot is placed at this point. Proceeding in the same way, the points are located for each age level.

4. Finally the dots or points are connected with the solid line that you see in the illustration (Figure 82).

### Interpretation

What can we learn from the line graph in Figure 82?

1. Between what two age levels do we show the least weight gain?
2. Between what two age levels do we show the greatest weight gain?
3. Does weight increase at a steady rate with age?
4. Is the *rate* of weight gain greater or lower from 8 to 11 than from 11 to 14?
5. What is the average annual weight gain per year from age 8 to 14?

### Using Two Line Graphs for Comparisons

Plotting two lines on the same graph is a common practice for showing significant comparisons.

In Figure 83 below we have plotted our original graph of average weights and the actual weights for a neighbor's son (Lenny) when he was in this age range.

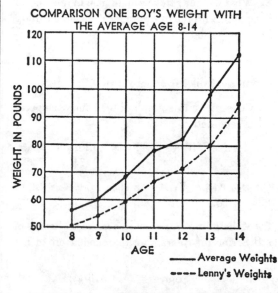

FIGURE 83

| Age | Lenny's Weight |
|-----|-----|
| 8 | 50 lb. |
| 9 | 53 lb. |
| 10 | 59 lb. |
| 11 | 67 lb. |
| 12 | 71 lb. |
| 13 | 80 lb. |
| 14 | 95 lb. |

### Practice Exercise No. 118

What can we learn from the comparison graphs in Figure 83?

**1.** At what age level was Lenny most below average?

**2.** At what age level was the boy least below average?

**3.** Between what two age levels did Lenny show the greatest weight gain?

**4.** Is this the same as occurred in the weight curve for the average group? What was the age of greatest gain in the average group?

**5.** What was Lenny's average annual weight gain per year from age 8 to 14?

**6.** By how much was Lenny's average annual weight gain greater or lower than that of the group average?

**7.** Comparing age 8 to 11 with age 11 to 14, in which period did Lenny show the greater weight gain?

**8.** What was his *rate* of weight gain in this period as compared with the *rate* gain for the average group during this period?

### Suggestions for Home Study Projects

**1.** Construct a line graph of the hourly temperatures as recorded:

| 8 A.M. | 9 A.M. | 10 A.M. | 11 A.M. | 12 noon | 1 P.M. |
|-----|-----|-----|-----|-----|-----|
| 55° | 58° | 62° | 70° | 74° | 76° |

| 2 P.M. | 3 P.M. | 4 P.M. | 5 P.M. | 6 P.M. | 7 P.M. |
|-----|-----|-----|-----|-----|-----|
| 78° | 76° | 75° | 73° | 71° | 68° |

**2.** Make a line graph of the population figures of the U.S. for the years shown, rounded off to the nearest million:

| 1979 | 1980 | 1981 |
|-----|-----|-----|
| 210,000,000 | 226,000,000 | 229,000,000 |

| 1982 | 1983 | 1984 |
|-----|-----|-----|
| 232,000,000 | 234,000,000 | 236,000,000 |

### PICTOGRAPHS

A picture graph or **pictograph,** as it is called, is a variation of the bar graph, using pictures or symbols instead of a bar. Generally the items pictured are associated with the subject of the comparison in the graph. Each picture or symbol represents a given quantity of the items being compared or illustrated.

In Figure 84 you see an example of a pictograph showing a comparison of the size of the populations of the Arab countries of the Middle East.

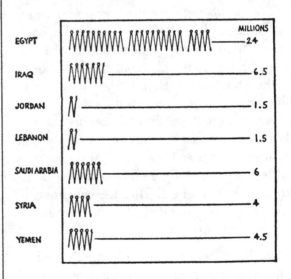

FIGURE 84. Each symbol represents 1 million inhabitants.

Do you see how the pictograph above could easily be converted to a bar graph?

Now look at the pictograph in Figure 85. Here a certain amount of humor has been introduced. Although it characteristically shows a comparison between stages of drunkenness and quantity of alcohol, it also provides a vivid pictorialization of the facts. Could we convert this to a bar graph? The answer is yes. Try it on your own. It will make an interesting exercise.

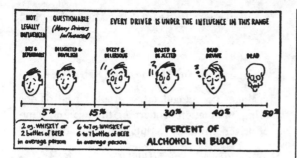

FIGURE 85. Diagram of the probable effect of certain percentages of alcohol in the blood.

You may notice that pictographs are frequently used to present statistics on such subjects as health, disease, and safety education. The purpose is to brighten up certain morbid, depressing facts which must be brought to the attention of the populace, even though they are unpleasant.

Figure 86 contains another type of pictograph showing the relation between the stopping distances of an automobile and stages of fatigue. The symbols include the picture of the automobile, and the distances on a football field. These make possible a comparison of the distance it takes to stop a car moving at 60 miles an hour in relation to the distances on a football field.

Many companies and agencies prefer to present information by means of pictographs because of the added interest and attention they attract.

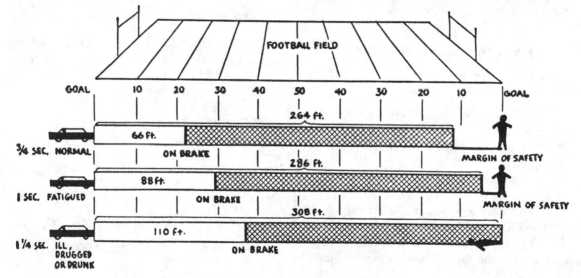

FIGURE 86. Figures based on tests of cars having brakes in first-class condition traveling at 60 MPH on dry level concrete surfaces.

### Suggestions for Home Study Practice

**1.** Make a pictograph of five different food items and their caloric value. You can get this information from any diet book.

**2.** Look in the classified advertisements of your local newspaper and make a pictograph that will compare the prices of several items.

### RECTANGLE GRAPHS

A **rectangle graph** is used when we wish to illustrate proportionate parts of quantities.

The graph consists simply of a rectangle with the indicated divisions.

The title usually gives the entire quantity, while each part is labeled in terms of what it represents. There are labeled axes such as those which appear in bar and line graphs.

The rectangle graph is especially useful when we wish to compare parts with each other or parts with the whole quantity.

You will sometimes see the rectangle graph (Figure 87) referred to as a single bar graph or divided bar graph. The reason for this is that it generally consists of one large single bar, divided into parts. Like the bar graph, it may run horizontally or vertically.

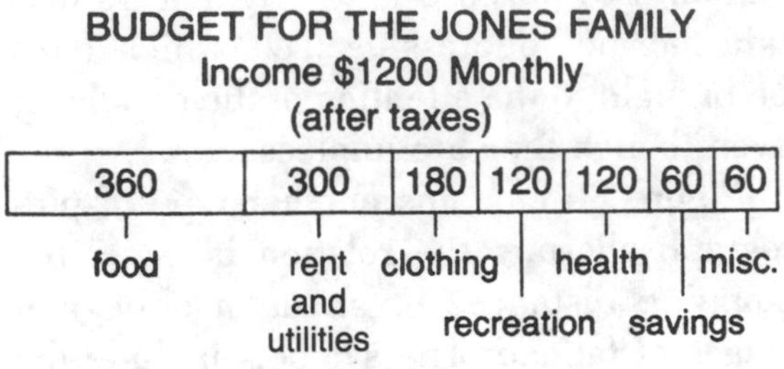

## BUDGET FOR THE JONES FAMILY
### Income $1200 Monthly
### (after taxes)

| 360 | 300 | 180 | 120 | 120 | 60 | 60 |
|-----|-----|-----|-----|-----|----|----|

food     rent   clothing   health    misc.
and
utilities      recreation   savings

FIGURE 87

In the rectangle graph you will note that the bar is divided into equal fractional parts. Each item takes up its share of the whole in such a way that relationships can be easily seen.

### Practice Exercise No. 118B

Answer the questions below by studying Figure 87.

**1.** Which two items account for more than half the budget?
**2.** What percentage of the income is set aside for clothing?
**3.** How much will the Jones family save in a year?
**4.** Which item consumes the largest part of the budget and how much does it amount to annually?
**5.** What is the ratio of the recreation item as compared with the expenditures for rent and utilities?

### *Hints for Constructing Rectangular or Divided Bar Graphs*

1. Find the total amount to be represented by the graph.
2. Find what fractional part of the total each item represents.
3. Convert the fractional parts to percents.
4. Divide the entire rectangle into equal units so that the fractional parts may be measured out easily.
5. Draw lines to show the percentage parts represented by each item and label the parts accordingly.

6. Write a title describing the subject of the graph.

### Suggestions for Home Study Practice

**1.** Make a rectangular bar graph based on the following information concerning the age and number of drivers involved in automobile accidents.

| Age | Number | % |
|-----|--------|---|
| Under 18 | 600,000 | ____ |
| 18–20 | 1,550,000 | ____ |
| 21–24 | 2,250,000 | ____ |
| 25–44 | 9,000,000 | ____ |
| 45–64 | 3,300,000 | ____ |
| 65 and over | 700,000 | ____ |

**2.** Make a rectangular graph based on the following facts relative to the Classifications of Expenditures of the Federal Government for the fiscal year 1952 (rounded to the nearest $\frac{1}{4}$ billion).

| Major Classification | Amount | % |
|----------------------|--------|---|
| National Defense | 39,000,000,000 | ____ |
| International Finance | 4,500,000,000 | ____ |
| Veterans | 5,250,000,000 | ____ |
| Interest on Public Debt | 6,000,000,000 | ____ |
| All Other | 11,250,000,000 | ____ |

## CIRCLE GRAPHS

The **circle graph** is used in the very same situations as the rectangular bar graph. It is used mostly to show how whole quantities are divided into proportionate parts.

A very popular use of circle graphs is to show how corporations, townships and government bodies apportion their budgets.

In constructing or using the circle graph to present a picture, the entire circle represents the whole. The parts, which are called sectors, are measured out as proportionate angles of the circle. As in the rectangle graph, each part is figured either as a fraction or percent of the whole.

We might best illustrate the similar application of the rectangular bar graph and the circle graph by converting the picture of the Jones family budget into a circle graph (Figure 88).

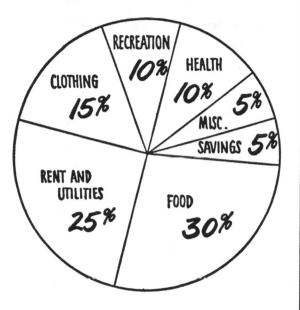

FIGURE 88

| Item | % of whole | Degrees |
|------|------------|---------|
| Food $360 | 30 | 108 |
| Rent $300 | 25 | 90 |
| Clothing $180 | 15 | 54 |
| Recreation $120 | 10 | 36 |
| Health $120 | 10 | 36 |
| Miscellaneous $60 | 5 | 18 |
| Savings $60 | 5 | 18 |

You can see from the table that it was necessary to convert the percentage represented by each item into degrees or portions of the circle in order to make the circle graph. What is the total number of degrees in a circle?

In the example below (Figure 89), you see a typical use of a circle graph by a corporation in reporting to the stockholders the way in which each dollar of income was spent.

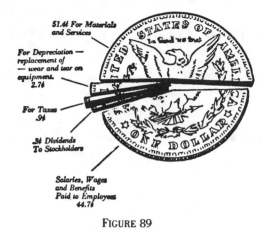

FIGURE 89

Consider what we learn from this graph almost at a glance.

1. Which item takes the largest part of the income?

2. What other single item takes close to half of the company's income?

3. What portion of the income is needed for replacement of worn equipment?

4. What part of the income goes to pay taxes?

5. What part do the stockholders get?

**Practice Exercise No. 119**

Your family decided to take a week's vacation and planned to spend $1250. In deciding on the budget, it was agreed that food would take $500, lodging $375, car expense $75, entertainment $175 and miscellaneous expenditures $125.

**1.** If you drew a circle graph of this budget, how many degrees would the sector on food take?

**2.** How many degrees would there be in the smallest sector? (To the nearest whole degree.)

**3.** What two sectors side by side could be shown as half the circle?

**4.** What fractional part of the expenses was to be spent on lodging?

**5.** What fractional part of the expense was allocated for entertainment?

**Suggestions for Home Study Practice**

**1.** Construct a circle graph to show the following facts about the Federal Government's sources of income.

| Source | Percentage | Type of Employment | Percentage of People |
|---|---|---|---|
| Income Taxes from Individuals | 40% | Manufacturing | 30% |
| Income Taxes from Corporations | 25% | Merchants | 25% |
| Customs and Other Import Taxes | 5% | Transportation | 5% |
| Excise Taxes | 20% | Professional Services | 10% |
| Borrowing | 10% | Personal Services | 12% |
| | | Others | 18% |

**2.** Construct a circle graph to show the following facts about the ways in which the residents of one city earned the majority of their income.

# CHAPTER FIFTEEN

# SIGNED NUMBERS

Jack and his sister Carolyn were playing shuffleboard. They both scored plus 7's. Then Jack landed in the "10 off" box. The score now stood $+7$ for Carolyn and $-3$ for Jack.

This is one of many instances in which it is helpful to use a plus sign or a minus sign in front of a number to indicate its direction and value.

Up to the present, all the numbers used in this book have been positive numbers. That is, none was less than zero (0). Although we did use the minus sign, it meant subtraction and did not represent a value of less than zero.

In solving some problems in arithmetic by short-cut methods, it is necessary to assign a negative value to some numbers. This is used primarily for numbers with which we desire to represent *opposite* quantities or qualities, and can best be illustrated by use of a diagram. For example consider a thermometer, as in Figure 90.

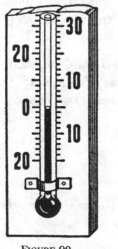

FIGURE 90

If temperatures above zero are taken as positive, then temperatures below zero are considered negative. Thus ten degrees above zero is written $+10°$ and ten degrees below zero would be written $-10°$.

BY DEFINITION: Numbers that have the $+$ or $-$ sign in front of them are called **signed numbers.**

## APPLYING SIGNED NUMBERS FOR OPPOSITE DIRECTIONS

In measuring distances east and west of a point, the use of signed numbers is applicable. For example, in referring to longitude, distances to the east of the zero meridian are designated as plus longitude and distances to the west as minus longitude. Thus, the Island of Sardinia, to the east would be designated as being in $+10°$ longitude and the City of Portugal, Spain, to the west, as being in $-10°$ longitude. Both these locations, one to the east and one to the west, are the same distance from the zero meridian.

*Latitude*—The Equator is taken as zero. North is plus and South is minus.

*Altitude*—Sea level is considered zero altitude. Points below sea level are minus, points above are plus.

*Bookkeeping*—Assets and money in the bank are plus quantities. Money owed or debits are minus amounts.

*Stock Market*—Daily changes in prices of stock are noted by plus and minus. A rise in price from the previous day's closing price is indicated as $+$, while a drop in price is shown as $-$.

You can see from these examples that signed numbers can be used in many different situations. They are vital when we wish to indicate the *direction* of a process as well as a number.

### Practice Exercise No. 120

Transpose the statements and numerical values into signed numbers.

1. 5 lb. overweight

2. 15° above zero temperature

3. 10 yard loss

4. 5% net loss

5. $50 profit

6. 32° S. latitude

7. 1000 ft. below sea level

8. 1858 A.D.

9. West longitude 55°

10. 18° below zero temperature

## ADDITION OF SIGNED NUMBERS

Learning to use signed numbers requires that you be introduced to some of the special rules employed in the study of algebra. It also prepares you for the equation method of solving some difficult arithmetic problems by an easier process.

In using signed numbers, keep in mind that the $+$ and $-$ continue to be used as signs of addition and subtraction as well as signs of positive and negative values. Since a positive number is the same as the numbers used in arithmetic, when no sign is indicated, the $+$ sign is understood.

EXAMPLE 1: A thermometer records a rise of 5° in one hour ($+5°$). An hour later it rises 4° more ($+4°$). What was its change in the two hours?

METHOD: $(+5°) + (+4°) = +9°$. Add the increases and the sum is given the plus sign, to show that the temperature went up.

EXAMPLE 2: The thermometer dropped 5° in one hour ($-5°$). An hour later it dropped another 4° ($-4°$). What was the temperature change in the two hours?

METHOD: $(-5°) + (-4°) = -9°$. Add the changes. The sum is 9 and the sign is $-$ because both changes were in the same direction.

**Rule: To add signed numbers of like signs,** *find the sum of the numbers and give it the common sign.*

EXAMPLE 3: A thermometer advanced 7 degrees ($+7°$) in one hour. The next hour it dropped 3 degrees ($-3°$). What was its change in this 2-hour period?

METHOD: Look at the thermometer in Figure 90. Place your finger on zero, go up 7 units, then go down 3 units. The result shows your finger at 4.

$$(+7°) + (-3°) = +4°$$

This is obtained by *subtracting the smaller from the larger*, and giving the result the *sign* of the *larger* quantity.

EXAMPLE 4: Suppose the thermometer advanced 7 degrees ($+7°$) the first hour and dropped 13 degrees ($-13°$) an hour later. What would be the total change?

METHOD: $(+7°) + (-13°) = -6°$.
This is obtained by subtracting the smaller from the larger and giving the result the sign of the larger quantity. Check the answer by tracing the steps on the thermometer.

**Rule: To add signed numbers of unlike signs,** *find the difference and give it the sign of the larger number.*

### Practice Exercise No. 121

Add the following signed numbers.

| | | | | | |
|---|---|---|---|---|---|
| 1. | +5<br>+3 | 3. | −5<br>−2 | 5. | +5<br>−2 |
| 2. | +4<br>−1 | 4. | −3<br>−5 | 6. | +13<br>− 6 |

| | | |
|---|---|---|
| **7.** $-\ 9$ $+15$ | **10.** $-22$ $+\ 6$ | **13.** $+19$ $-\ 9$ $-\ 4$ |
| **8.** $-18$ $+\ 7$ | **11.** $+12$ $+\ 5$ $+\ 3$ | **14.** $-5$ $+6$ $-8$ |
| **9.** $+\ 9$ $-14$ | **12.** $-1$ $-3$ $+8$ | **15.** $-17$ $+14$ $-\ 6$ |

## SUBTRACTION OF SIGNED NUMBERS

In subtracting signed numbers, we are finding the *difference* between two values on a scale.

Using the thermometer again for illustration, if we asked what is the difference in degrees between $-6°$ and $+5°$, your answer would be $11°$. You can easily do this mentally.

Now we ask, what method did you use to arrive at the answer? First you counted from $-6°$ to zero, then added 5 to that. From this procedure we can derive the following rule for subtraction of signed numbers.

**Rule: To subtract signed numbers,** *change the sign of the subtrahend and apply the rules for addition.*

EXAMPLE 1: What is the difference between a point 30 miles below the equator and a location 18 miles directly north of the equator.

METHOD: Subtract $-30$ from $+18$. $-30$ is the subtrahend or number to be subtracted.

When we change its sign and add, we get:

$$+30\ +\ 18\ =\ 48 \text{ miles Ans.}$$

EXAMPLE 2: From $-26$ subtract $-14$.

METHOD: $-14$ is the subtrahend. When we change its sign and add, we get

$$-26\ +\ 14\ =\ -12 \text{ Ans.}$$

**Practice Exercise No. 122**

The exercise below will test your ability to subtract plus and minus quantities. *Subtract:*

| | | |
|---|---|---|
| **1.** $+14$ $+\ 8$ | **4.** $+12$ $+11$ | **7.** $-58$ $-72$ |
| **2.** $+8$ $-3$ | **5.** $-17$ $-13$ | **8.** $+77$ $-22$ |
| **3.** $-9$ $+5$ | **6.** $-\ 3$ $-16$ | **9.** $(-6) - (-9)$ |
| | | **10.** $(-8) - (+10)$ |

## ALGEBRAIC SUMS AND DIFFERENCES

In carrying out the addition and subtraction of signed numbers, you have been finding what is called *algebraic sums and differences*.

Using the identical procedures you can add or subtract numbers that are represented by symbols.

EXAMPLE 1: Add $-6a\ +\ 13a\ +\ 7a$.

METHOD:
$$13a\ +\ 7a\ =\ 20a;$$
$$20a\ -\ 6a\ =\ 14a \text{ Ans.}$$

EXAMPLE 2: Subtract $3a$ from $8a$.

METHOD: $8a\ -\ 3a\ =\ 5a$ Ans.

Here we are working with *like* terms. We cannot add or subtract *unlike* terms. For instance, if we let $a$ stand for apples and $b$ stand for books, we know from our fundamentals of arithmetic that we could not combine apples and books into a single quantity of either. Therefore, **to add or subtract quantities containing unlike symbols,** collect like terms and express them separately in the answer.

EXAMPLE 3: Add $6a + 4b + 3b + 2a$.

METHOD: Collecting like terms,
$$6a + 2a = 8a$$
$$4b + 3b = 7b$$

Expressing unlike terms separately, we get $8a + 7b$ ANS.

This is an algebraic expression containing two terms.

### Practice Exercise No. 123

The following exercise will test your knowledge of addition and subtraction of signed numbers.

*Add*

1. $6 + 5 + 4 - 3 =$
2. $-4 - 16 - 13 =$
3. $7 - 14 - 21 + 3 =$
4. $-12a - 3a =$
5. $8c - 15c =$
6. $9b - 4b - 3b =$
7. $3a + 5a + 7a - 2b =$
8. $7a - 3b + 4a - 5b =$

*Subtract*

9. $(19) - (6) =$
10. $(18) - (-4) =$
11. $(-31) - (5) =$
12. $(-8) - (19) =$
13. $(-11) - (-41) =$
14. $(5a) - (2a) =$
15. $(-18b) - (7b) =$
16. $(17c) - (19c) =$

# CHAPTER SIXTEEN

# HAND-HELD CALCULATOR

The first calculating machine was probably a pile of sticks or stones, and the person doing the calculating no doubt did so by adding or subtracting from the pile. The **abacus** is a hand-operated calculating machine in which numbers are represented by beads on wires. The beads are in columns, contained in a rectangular frame, and are moved up and down to do the calculating. The first abacus was no doubt sand in which marks were made and rubbed out as required. The most common form of the abacus is the ancient Chinese version called the *suan pan*, which means reckoning board. It has up to 13 columns of beads, with each column divided into 2 parts. The upper columns have 2 beads each and are separated from the lower columns by a horizontal bar. The lower columns have 5 beads each (see Figure 91).

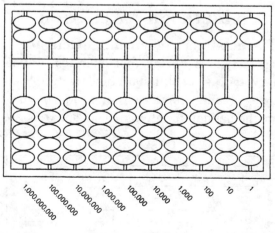

FIGURE 91

In the first column on the right the lower beads represent 1 each, while the upper represent 5. The next column to the left is the tens' column, and each lower bead represents 10 and each upper bead 50. In the next column the lower beads represent 100 each and the upper beads 500 each, and so on across the abacus. To begin use of the abacus all the lower beads are at the bottom of the frame and all the upper beads are at the top of the frame (as in Figure 91). The abacus pictured has only ten columns and with just these 70 beads numbers up to 9,999,999,999 can be represented. Numbers are entered by moving beads against the crossbar. The number 32 is entered by moving three beads in the tens' column (second) against the crossbar and two beads against the crossbar in the ones' column (first). See Figure 92.

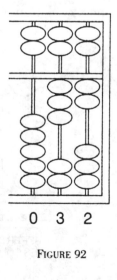

0　3　2

FIGURE 92

To accomplish addition, 32 + 6, we move one upper bead in the ones' column down to the crossbar (thus adding five) and one lower bead is slid up toward the crossbar (adding one more, making six—see Figure 93). The abacus now reads 38.

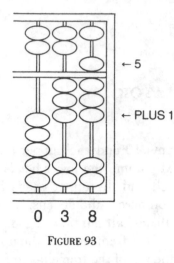

← 5

← PLUS 1

0    3    8

FIGURE 93

Now to add 2 more to the total of 38 (38 + 2 = 40) by moving the last two beads in the ones column up to the crossbar. The lower part of the column is now full (Figure 94A), and whenever this happens it must be emptied immediately. This is done by moving the last bead in the upper part of the ones' column against the bar, adding five, and moving the five lower beads down to the bottom of the frame. But now the upper part of the ones' column is

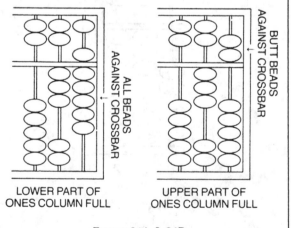

LOWER PART OF
ONES COLUMN FULL

UPPER PART OF
ONES COLUMN FULL

FIGURE 94A & 94B

"full" and must be emptied. (See Figure 94B.) This is done by moving one more bead in the tens' column up against the bar and then subtracting the ten in the ones' column by moving the two upper beads to the top of the frame (see Figure 95).

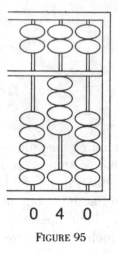

0    4    0

FIGURE 95

Of course, a skilled operator would foresee many of the steps and skip them. With skill it is possible to calculate at very high speeds on an abacus. As a matter of fact some people can use an abacus faster than a skilled person can use an electric calculator. There are many books on the abacus if you want to learn how to use one with speed and accuracy.

Subtraction is performed in much the same way, except that beads are moved away from the crossbar instead of toward it, as in addition.

## ELECTRONIC CALCULATOR

The **hand-held electronic calculator** is an ingenious device that operates by battery and uses solid-state devices like transistors and logic circuits to perform mathematical functions. The answers are

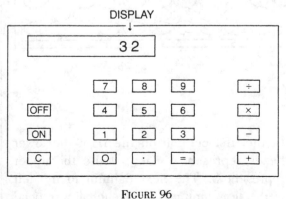

FIGURE 96

usually read on a display that is made up of **LCDs (liquid crystal diodes)** or **LEDs (light-emitting diodes)**. LCD displays are black on a gray background, while LED displays are usually red on black. Data is entered via a keyboard like the one shown in Figure 96.

When numbers are entered, they begin to fill the display by entering from right to left. If the number 3 is entered, it appears in the right-most position of the display. If we then enter the number 2, the display will read 32. Numbers will continue to move across the display until a *function* key is pressed ($+$, $\times$, $-$, $\div$). When a function key is pressed, the number that is in the display and the function are stored by the machine. If we wish to add 32 to the number 6, we would enter 3, then 2, (the display now reads 32); then we press the $+$ key (the display still reads 32, but the value 32 is also stored *inside* the machine); now we press the number 6 (the display now reads 6) and the storage *inside* the machine now also contains the number 6 as well as the function $+$. Now when we press the "$=$" key, the machine adds the 6 and the 32, and the number 38 appears in the display. If we now wish to add 2 to the previous result of 38, we press the $+$ and the 38 enters into storage; now pressing 2 will cause the 2 to appear in the display, and the $+$ and 2 are stored. Now pressing $=$ causes the addition to take place and the display reads 40.

We can continue to add to this number by repeating the steps above. However, it is not necessary to press the "$=$" key each time, only at the last step when we wish to see the answer that is stored.

To clarify things let's do some examples:

EXAMPLE: Add 32 to 40 to 76.

METHOD:

Enter 32   press $+$   enter 40   press $+$ enter 76   press $=$

The display reads 148. ANS.

EXAMPLE: Add 456 to 32 to 11 and subtract 41.

METHOD:

Enter 456   press $+$   enter 32   press $+$ enter 11   press $-$   enter 41   press $=$ The display reads 458. ANS.

When we wish to enter a number that has a decimal part or is a decimal number, we use the key marked with the decimal point, ".".

EXAMPLE: Add 3.2 to .05 to 4.

METHOD:

Enter 3   press .   enter 2   press $+$ press .   enter 05   press $+$   enter 4 press $=$

The display reads 7.25. ANS.

Multiplication is performed by entering the multiplicand then pressing the function key marked "$\times$" followed by the multiplier, then press the key marked "$=$" to read the answer in the display.

EXAMPLE: Multiply $5 \times 3$

METHOD:

Enter 5   press $\times$   enter 3   press $=$ The display reads 15. ANS.

Division is performed in the same way. We enter the dividend, press the function key marked "$\div$", enter the divisor, then press the key marked "$=$" to read the quotient in the display.

EXAMPLE: Divide 73 by 12

METHOD:

Enter 73   press $\div$   enter 12   press $=$ The display reads 6.08333333 ANS.

The key marked "C" when pressed will erase the number that is in the display, so if we should enter a number by mistake, we press C and the display will read 0. What if in the previous example we enter 32, then the + function, but then the 7 when we wanted to add 6? If at this time we press the key marked "C" the 7 will be erased from the display, but the accumulator still has 32 in it even though we cannot see it. Now we again press the function key + (because this was erased also when we erased the 7) followed by 6, then =, and the display will read 38. Remember the clear "C" erases the function as well as the number, so the function must be reentered as well as the number. It is a good idea to press the C key before each new use of the calculator to be sure that there are no values remaining in the machine when we begin. This is not done in the examples only to save space in printing.

*Not all calculators work the same way* when we add or subtract and then multiply.

EXAMPLE: In one calculator when we:

enter 3   press +   enter 6   press ×
enter 5   press =
The display reads 33 ANS.

EXPLANATION: The calculation performed was 3 + (6 × 5) = 33 *(the multiplication is performed before the addition)*.

EXAMPLE: In a different calculator when we:

enter 3   press +   enter 6   press ×
enter 5   press =
The display reads 45 ANS.

EXPLANATION: The calculation performed was (3 + 6) × 5 = 45 *(the addition is performed before the multiplication)*.

Most calculators work as in the second example above, but the instruction booklet that comes with the calculator should make this clear.

Even the very inexpensive calculators can do more functions than just addition,

subtraction, multiplication, and division. (See Figure 97.)

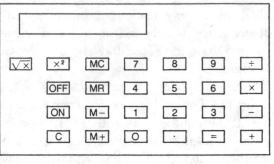

FIGURE 97

The keys marked, $X^2$, $\sqrt{X}$, MC, MR, M−, and M+ will now be explained.

### Key Marked "$X^2$"

This key when pressed will find the square of the display (the value multiplied by itself).

EXAMPLES: Enter 5   press $X^2$
The display reads 25. (5 × 5)

Enter 2.5   press $X^2$
The display reads 6.25 (2.5 × 2.5)

### Key Marked "$\sqrt{X}$"

This key when pressed will find the square root of the display (a value that when multiplied by itself will equal the display).

EXAMPLES: Enter 36   press $\sqrt{X}$
The display reads 6 (6 × 6 = 36)

Enter 6.25   press $\sqrt{X}$
The display reads 2.5 (2.5 × 2.5 = 6.25)

Enter 4568   press $\sqrt{X}$
The display reads 67.8698 (67.8698 × 67.8698 = 4568)

### Keys Marked "MC," "MR," "M−," and "M+"

These keys refer to a register in the machine called the **memory.** The memory is like the display, but it cannot be seen and is a place where values can be stored.

### Key Marked "MC"

This key is like the clear key, but it clears the register called the memory and leaves the display as it was, use it to be sure the memory register is clear before using.

### Key Marked "MR"

This key takes the value from the memory and puts it in the display.

### Key Marked "M –"

This key will subtract the value of the display from the value of the memory; the value of the display is unchanged while the value of the memory is decreased.

### Key Marked "M +"

This key will add the value of the display to the value of the memory; the value of the display will be unchanged while the value of the memory is increased.

EXAMPLE:

$$(23 \times 7) + (12 \div 5) - (27.34 + .165)$$

METHOD:

| Action | Display reads | Memory contains |
|---|---|---|
| Press MC | 0 | 0 |
| Enter 23   press × <br> enter 7   press M + | 161 | 161 |
| Enter 12   press ÷ <br> enter 5   press M + | 2.4 | 163.4 |
| Enter 27.34   press + <br> enter .165   press M − | 27.505 | 135.895 |
| Press MR | 135.895 | 135.895 |

Notice that the display contains the answer; also notice that the memory still has the value 135.895 and *will until we change it.*

There are hand-held calculators that are really computers. That is, they can be set up with a set of instructions that will do a sequence of mathematical operations once they are *programmed*, but these require a knowledge of algebra.

You have spent a great deal of time learning arithmetic and becoming proficient with addition, multiplication, division, and subtraction, and should be congratulated. One word of caution: don't depend too much on calculators or your proficiency will suffer.

# ANSWERS

## Exercise No. 1

| | | |
|---|---|---|
| **1.** VIII | **7.** LXXVI | **12.** DCCCII |
| **2.** XVI | **8.** LXXXIX | **13.** M̲CMLVIII |
| **3.** XXIV | **9.** XCI | **14.** X̄CC |
| **4.** XXXIX | **10.** CXIV | **15.** C̲ |
| **5.** XLVIII | **11.** CDLVI | **16.** M̄M |
| **6.** LIII | | |

## Exercise No. 2

| | | | |
|---|---|---|---|
| **1.** 7 | **5.** 78 | **9.** 360 | **13.** 1960 |
| **2.** 23 | **6.** 92 | **10.** 631 | **14.** 10,300 |
| **3.** 46 | **7.** 105 | **11.** 971 | **15.** 150,020 |
| **4.** 69 | **8.** 215 | **12.** 1002 | **16.** 1,905,000 |

## Exercise No. 3

| | |
|---|---|
| **1.** 698 | **6.** 462,309 |
| **2.** 2465 | **7.** 6,422,754 |
| **3.** 3412 | **8.** 9,002,000,050 |
| **4.** 33,600 | **9.** 604,368,419 |
| **5.** 301,465 | **10.** 21,000,000,400 |

## Exercise No. 4

| | | |
|---|---|---|
| **1.** 390 | **7.** 82,700 | **13.** 30,000 |
| **2.** 4620 | **8.** 96,300 | **14.** 70,000 |
| **3.** 680 | **9.** 26,000 | **15.** 390,000 |
| **4.** 8240 | **10.** 69,000 | **16.** 5,400,000 |
| **5.** 700 | **11.** 389,000 | |
| **6.** 6300 | **12.** 5,395,000 | |

## Exercise No. 5

(a) 368   (b) 401   (c) 390   (d) 413   (e) 405

## Exercise No. 6

| | | |
|---|---|---|
| (a) 3231 | (c) 23,429 | (e) 136,848 |
| (b) 4029 | (d) 11,249 | |

## Exercise No. 7

| | | | | |
|---|---|---|---|---|
| **1.** 76 | **6.** 47 | **11.** 59 | **16.** 47 | **21.** 53 |
| **2.** 54 | **7.** 62 | **12.** 66 | **17.** 74 | **22.** 82 |
| **3.** 59 | **8.** 60 | **13.** 60 | **18.** 40 | **23.** 80 |
| **4.** 51 | **9.** 85 | **14.** 92 | **19.** 91 | **24.** 72 |
| **5.** 68 | **10.** 61 | **15.** 81 | **20.** 112 | **25.** 114 |

## Exercise No. 8

| | | | | |
|---|---|---|---|---|
| **1.** 182 | **3.** 187 | **5.** 268 | **7.** 1884 | **9.** 23,896 |
| **2.** 272 | **4.** 191 | **6.** 2248 | **8.** 3947 | **10.** 28,055 |

## Exercise No. 9

| **1.** | **2.** | **3.** | **4.** | **5.** |
|---|---|---|---|---|
| 39 | 33 | 37 | 47 | 31 |
| 40 | 42 | 41 | 31 | 37 |
| 39 | 43 | 30 | 33 | 32 |
| 33 | 34 | 40 | 38 | 35 |
| 37,339 | 38,753 | 43,447 | 41,657 | 38,601 |

## Exercise No. 10

| | | | | |
|---|---|---|---|---|
| **1.** (a) 215 | (b) 136 | (c) 484 | (d) 293 | (e) 399 |
| **2.** (a) 52 | (b) 286 | (c) 85 | (d) 284 | (e) 39 |
| **3.** (a) 313 | (b) 123 | (c) 398 | (d) 56 | (e) 2524 |
| **4.** (a) 4410 | (b) 5525 | (c) 2201 | (d) 3018 | (e) 37982 |

## Exercise No. 11

| | |
|---|---|
| **1.** $.12 | **6.** $14.05 |
| **2.** $.06 | **7.** $220.00 |
| **3.** $.60 | **8.** $2,400.35 |
| **4.** $1.01 | **9.** $12,684.19 |
| **5.** $1.32 | **10.** $3,000,030.98 |

## Exercise No. 12

| | |
|---|---|
| **1.** $73.28 | **6.** $6.68 |
| **2.** $45.47 | **7.** $2.15 |
| **3.** $6.20 | **8.** $2.99 |
| **4.** $24.34 | **9.** $9.69 |
| **5.** $331.29 | **10.** $337.59 |

## Exercise No. 13

| | | |
|---|---|---|
| **1.** $3.45 | **3.** $4.45 | **5.** $6780.82 |
| **2.** $10.25 | **4.** $13.11 | |

## Exercise No. 14

| | | | | |
|---|---|---|---|---|
| **1.** (a) 129 | (b) 80 | (c) 224 | (d) 528 | (e) 96 |
| **2.** (a) 504 | (b) 320 | (c) 672 | (d) 783 | (e) 360 |
| **3.** (a) 2884 | (b) 2196 | (c) 4536 | (d) 1224 | (e) 5250 |

## Exercise No. 15

| | | | |
|---|---|---|---|
| **1.** 200 | **4.** 900 | **7.** 1600 | **10.** 2000 |
| **2.** 360 | **5.** 1500 | **8.** 6000 | **11.** 24,000 |
| **3.** 2000 | **6.** 4200 | **9.** 9300 | **12.** 120,000 |

## Exercise No. 16

| | | |
|---|---|---|
| **1.** 27,608 | **6.** 80,275 | **11.** 274,176 |
| **2.** 307,098 | **7.** 60,775 | **12.** 154,635 |
| **3.** 36,284 | **8.** 205,227 | **13.** 323,680 |
| **4.** 35,108 | **9.** 410,112 | **14.** 248,920 |
| **5.** 26,643 | **10.** 452,226 | **15.** 550,854 |

## Exercise No. 17

| | | |
|---|---|---|
| **1.** $29.76 | **5.** $22.00 | **9.** $39,823.00 |
| **2.** $118.56 | **6.** $1972.08 | **10.** $11,222.60 |
| **3.** $322.56 | **7.** $130.00 | **11.** $67,210.90 |
| **4.** $710.60 | **8.** $499.00 | **12.** $542,406.00 |

## Exercise No. 18

| | | | | |
|---|---|---|---|---|
| **1.** 147 | **4.** 288 | **7.** 4032 | **10.** 4256 | **13.** 810 |
| **2.** 192 | **5.** 195 | **8.** 3672 | **11.** 288 | **14.** 559 |
| **3.** 333 | **6.** 2142 | **9.** 4854 | **12.** 465 | **15.** 576 |

## Exercise No. 19

| | | | |
|---|---|---|---|
| **1.** 335 | **8.** 913 | **15.** 7128 | **22.** 33,904 |
| **2.** 365 | **9.** 1666 | **16.** 8484 | **23.** 26,574 |
| **3.** 774 | **10.** 2244 | **17.** 7171 | **24.** 41,031 |
| **4.** 846 | **11.** 3984 | **18.** 25,344 | **25.** 33,792 |
| **5.** 702 | **12.** 3224 | **19.** 28,017 | |
| **6.** 616 | **13.** 6732 | **20.** 24,442 | |
| **7.** 682 | **14.** 5643 | **21.** 15,049 | |

## Exercise No. 20

| | | | | | | | |
|---|---|---|---|---|---|---|---|
| 1 | 8 | 7 | 8 | 4 | 5 | 9 | 4 |
| 7 | 6 | 4 | 9 | 2 | 1 | 7 | 3 |
| 1 | 8 | 1 | 1 | 8 | 1 | 1 | 7 |
| 2 | 2 | 5 | 2 | 8 | 3 | 9 | 5 |
| 5 | 5 | 5 | 9 | 8 | 3 | 8 | 2 |
| 3 | 1 | 9 | 3 | 7 | 2 | 7 | 6 |
| 3 | 7 | 7 | 8 | 2 | 6 | 2 | 6 |
| 6 | 5 | 8 | 4 | 5 | 4 | 6 | 4 |
| 9 | 9 | 7 | 6 | 6 | 9 | 2 | 4 |
| 3 | 9 | 4 | 3 | 3 | 5 | 1 | 6 |

## Exercise No. 21

| | |
|---|---|
| **1.** 75 | **7.** 311 |
| **2.** 46 | **8.** 322 |
| **3.** 821 | **9.** 322 R2 |
| **4.** 511 | **10.** 22 R1 |
| **5.** 62 | **11.** 210 R1 |
| **6.** 29 | **12.** 1157 R5 |

## Exercise No. 22

| | | |
|---|---|---|
| **1.** 240 R10 | **8.** 47 R9 | **15.** 314 |
| **2.** 97 | **9.** 73 R26 | **16.** 340 |
| **3.** 23 | **10.** 946 R83 | **17.** 70 |
| **4.** 82 | **11.** 27 | **18.** 52 R66 |
| **5.** 310 | **12.** 43 | **19.** 57 R112 |
| **6.** 105 | **13.** 406 R2 | **20.** 74 R406 |
| **7.** 45 R15 | **14.** 104 | |

## Exercise No. 23

| | | | | |
|---|---|---|---|---|
| **1.** 8 | **4.** 5 | **7.** 8 | **10.** 0 | **13.** 0 |
| **2.** 6 | **5.** 1 | **8.** 2 | **11.** 3 | **14.** 8 |
| **3.** 0 | **6.** 0 | **9.** 6 | **12.** 4 | **15.** 3 |

## Exercise No. 24

| | |
|---|---|
| **1.** Right | **6.** Wrong |
| **2.** Right | **7.** Right |
| **3.** Wrong | **8.** Right |
| **4.** Wrong | **9.** Wrong |
| **5.** Right | **10.** Wrong |

## Exercise No. 25

| | |
|---|---|
| **1.** 1820 | **6.** B |
| **2.** $3750.00 | **7.** B |
| **3.** $462.00 | **8.** A |
| **4.** 1152 Tomatoes | **9.** C |
| **5.** $717,600.00 | **10.** C |

## Exercise No. 26

| | | |
|---|---|---|
| **1.** 685 R4 | **5.** 720 R6 | **9.** 2164 R6 |
| **2.** 765 R4 | **6.** 16986 R1 | **10.** 3403 |
| **3.** 1019 R2 | **7.** 11301 R4 | |
| **4.** 456 R5 | **8.** 2222 R2 | |

## Exercise No. 27

| | | | |
|---|---|---|---|
| **1.** $.58 | **3.** $.92 | **5.** $2.65 | **7.** $.48 |
| **2.** $7.93 R83 | **4.** $5.89 | **6.** $7.98 | **8.** $.93 |

## Exercise No. 28

**1.** Subtract
$2852.67
− 450.00
$2402.67

**2.** Divide
$ 6.50
64 ) $416.00

**3.** Add
3455
4823
3237
3684
15,199

**4.** Multipy
267
$7.80
21360
1867
$2082.60

**5.** Add—Subtract
283
334
247
970
1834

2000
− 1834
166 Ans.

**Exercise No. 28** *(Continued)*

**6.** Divide

```
        183
56 ) 10248
     56
     464
     448
     168
     168
       0
```

**7.** Divide—Multiply

```
      732
14 ) 10248

         732
       × $.89
       6588
       5856
     $651.48 Ans.
```

**8.** Multiply
(a) 6 × 500 = 3000 ft.
    Not enough
(b) Divide

```
      416 pkgs.
6 ) 2500
```

**9.** Add and Divide

```
  289
  269
  246
  804
```

```
           268 Ans.
3 ) 804
```

**10.** (a) Multiply 165 × $91 = 15,015
(b) Add 91 + 8 = $99
    Multiply 157 × 99 = 15,543
(c) Subtract 15543
              − 15015
                 528

### Exercise No. 29

**1.** $\frac{3}{9}$  **4.** $\frac{15}{20}$  **7.** $\frac{4}{14}$  **10.** $\frac{4}{48}$  **13.** $\frac{36}{81}$

**2.** $\frac{8}{12}$  **5.** $\frac{3}{18}$  **8.** $\frac{5}{25}$  **11.** $\frac{6}{33}$  **14.** $\frac{25}{35}$

**3.** $\frac{4}{16}$  **6.** $\frac{18}{36}$  **9.** $\frac{9}{24}$  **12.** $\frac{15}{60}$  **15.** $\frac{35}{42}$

### Exercise No. 30

**1.** $\frac{3}{8}$  **6.** $\frac{3}{5}$  **11.** $\frac{2}{5}$  **16.** $\frac{5}{9}$

**2.** $\frac{3}{8}$  **7.** $\frac{4}{25}$  **12.** $\frac{4}{9}$  **17.** $\frac{15}{32}$

**3.** $\frac{1}{4}$  **8.** $\frac{2}{3}$  **13.** $\frac{4}{5}$  **18.** $\frac{1}{3}$

**4.** $\frac{1}{2}$  **9.** $\frac{7}{8}$  **14.** $\frac{3}{14}$  **19.** $\frac{3}{8}$

**5.** $\frac{2}{3}$  **10.** $\frac{1}{3}$  **15.** $\frac{3}{10}$  **20.** $\frac{1}{2}$

### Exercise No. 31

**1.** $\frac{16}{3}$  **7.** $\frac{14}{3}$  **13.** $\frac{94}{7}$  **19.** 17  **25.** $2\frac{2}{3}$

**2.** $\frac{63}{5}$  **8.** $\frac{13}{9}$  **14.** $\frac{71}{5}$  **20.** $6\frac{1}{4}$  **26.** $5\frac{1}{4}$

**3.** $\frac{10}{3}$  **9.** $\frac{23}{4}$  **15.** $\frac{112}{5}$  **21.** 7  **27.** $7\frac{1}{2}$

**4.** $\frac{35}{8}$  **10.** $\frac{26}{3}$  **16.** $8\frac{1}{3}$  **22.** $3\frac{1}{2}$  **28.** $4\frac{1}{10}$

**5.** $\frac{73}{6}$  **11.** $\frac{97}{6}$  **17.** $12\frac{3}{5}$  **23.** 5  **29.** $4\frac{2}{11}$

**6.** $\frac{17}{7}$  **12.** $\frac{86}{7}$  **18.** $3\frac{1}{8}$  **24.** $1\frac{3}{5}$  **30.** $5\frac{1}{3}$

### Exercise No. 32

**1.** 12  **4.** 18  **7.** 24  **10.** 30

**2.** 10  **5.** 12  **8.** 8  **11.** 63

**3.** 15  **6.** 24  **9.** 24  **12.** 315

### Exercise No. 33

**1.** $1\frac{6}{7}$  **5.** 24  **9.** $17\frac{2}{3}$  **13.** $15\frac{17}{20}$

**2.** $6\frac{1}{2}$  **6.** $1\frac{1}{6}$  **10.** $18\frac{2}{3}$  **14.** $35\frac{1}{24}$

**3.** $5\frac{1}{2}$  **7.** $7\frac{1}{3}$  **11.** $12\frac{2}{3}$  **15.** $16\frac{7}{12}$

**4.** $14\frac{2}{3}$  **8.** $8\frac{7}{10}$  **12.** $24\frac{1}{8}$

### Exercise No. 34

**1.** $\frac{1}{2}$  **5.** $8\frac{1}{10}$  **9.** $9\frac{6}{7}$  **13.** $2\frac{5}{6}$

**2.** $\frac{1}{4}$  **6.** $9\frac{1}{6}$  **10.** $11\frac{1}{16}$  **14.** $4\frac{23}{24}$

**3.** $7\frac{1}{2}$  **7.** $7\frac{3}{4}$  **11.** $6\frac{1}{3}$  **15.** $3\frac{7}{12}$

**4.** $2\frac{1}{24}$  **8.** $\frac{1}{3}$  **12.** $4\frac{8}{9}$  **16.** $\frac{1}{8}$

### Exercise No. 35

**1.** $25\frac{7}{8}$  **6.** $16\frac{1}{2}$

**2.** $2\frac{1}{4}$ inches  **7.** (a) $3\frac{3}{4}$  (b) $3\frac{1}{2}$

**3.** 1 hour  **8.** $1\frac{7}{12}$ hours

**4.** 10 rolls  **9.** $1\frac{7}{8}$ pounds

**5.** $1\frac{1}{4}$  **10.** $11.25

### Exercise No. 36

**1.** 1  **5.** $2\frac{2}{3}$  **9.** $2\frac{6}{7}$  **13.** $1\frac{7}{8}$

**2.** $\frac{2}{3}$  **6.** $1\frac{3}{7}$  **10.** $1\frac{7}{8}$  **14.** $4\frac{1}{2}$

**3.** 1  **7.** $3\frac{1}{5}$  **11.** $1\frac{1}{8}$  **15.** $1\frac{7}{9}$

**4.** $12\frac{4}{5}$  **8.** 9  **12.** $1\frac{1}{3}$  **16.** $4\frac{2}{3}$

### Exercise No. 37

**1.** $4\frac{1}{2}$  **4.** 21  **7.** 58  **10.** 32

**2.** $2\frac{2}{3}$  **5.** 37  **8.** 256  **11.** 700

**3.** 50  **6.** $14\frac{1}{4}$  **9.** $171\frac{1}{2}$  **12.** 2072

### Exercise No. 38

**1.** $\frac{3}{16}$  **4.** $\frac{1}{4}$  **7.** $\frac{3}{4}$  **10.** $1\frac{3}{4}$

**2.** $\frac{1}{28}$  **5.** $\frac{11}{16}$  **8.** $\frac{7}{15}$  **11.** 9

**3.** $\frac{5}{18}$  **6.** $\frac{1}{24}$  **9.** 5  **12.** $5\frac{1}{2}$

### Exercise No. 39

**1.** $\frac{5}{24}$  **4.** $\frac{7}{32}$  **7.** 50  **10.** 64

**2.** $\frac{1}{9}$  **5.** $\frac{1}{21}$  **8.** 24  **11.** 20

**3.** $\frac{1}{6}$  **6.** 8  **9.** 49  **12.** 72

### Exercise No. 40

**1.** 4  **4.** $\frac{11}{14}$  **7.** $1\frac{10}{11}$  **10.** $16\frac{2}{3}$

**2.** $\frac{11}{20}$  **5.** $5\frac{3}{4}$  **8.** $7\frac{3}{7}$  **11.** $2\frac{2}{3}$

**3.** $6\frac{1}{2}$  **6.** $4\frac{1}{2}$  **9.** 34  **12.** $1\frac{5}{6}$

## Exercise No. 41

| | | | | |
|---|---|---|---|---|
| **1.** 45 | **5.** 25 | **9.** $60 | **13.** 48 | **17.** 52 |
| **2.** 84 | **6.** 24 | **10.** 18 gal | **14.** 28 | **18.** 4 |
| **3.** 13 | **7.** 36 | **11.** 6 | **15.** $\frac{1}{10}$ | **19.** 23 |
| **4.** 18 | **8.** 32 | **12.** 42¢ | **16.** 80 | **20.** 3 |

## Exercise No. 42

| | | |
|---|---|---|
| **1.** f | **6.** b | **11.** m |
| **2.** n | **7.** o | **12.** d |
| **3.** a | **8.** c | **13.** h |
| **4.** g | **9.** k | **14.** j |
| **5.** i | **10.** e | **15.** l |

## Exercise No. 43

| | |
|---|---|
| **1.** 5.7 | **6.** .2 |
| **2.** 15.28 | **7.** .005 |
| **3.** 42.006 | **8.** 4.04 |
| **4.** 223.3 | **9.** .052 |
| **5.** 9029.15 | **10.** .32 |

## Exercise No. 44

| | | | | |
|---|---|---|---|---|
| **1.** .5 | **3.** 5 | **5.** 5.6 | **7.** 53.001 | **9.** 2.91 |
| **2.** .42 | **4.** .3 | **6.** 1.1 | **8.** .0401 | **10.** .008 |

## Exercise No. 45

| | | | |
|---|---|---|---|
| **1.** $\frac{2}{5}$ | **6.** $\frac{1}{250}$ | **11.** .60 | **16.** .80 |
| **2.** $\frac{1}{20}$ | **7.** $\frac{41}{400}$ | **12.** .625 | **17.** .833 |
| **3.** $\frac{4}{25}$ | **8.** $\frac{1}{50000}$ | **13.** .313 | **18.** .875 |
| **4.** $\frac{7}{25}$ | **9.** $\frac{801}{1000}$ | **14.** .75 | **19.** .563 |
| **5.** $\frac{7}{10}$ | **10.** $\frac{90009}{100000}$ | **15.** .167 | **20.** .281 |

## Exercise No. 46

| | | | | |
|---|---|---|---|---|
| **1.** 0 | **3.** 4 | **5.** 1 | **7.** 0 | **9.** 1 |
| **2.** 3 | **4.** 1 | **6.** 1 | **8.** 4 | **10.** 2 |

## Exercise No. 47

| | |
|---|---|
| **1.** 2.3 | **6.** .230 |
| **2.** 1.76 | **7.** 10.019 |
| **3.** 13.7 | **8.** 129.94 |
| **4.** 2.88 | **9.** 22.1304 |
| **5.** 19.62 | **10.** 75.399 |

## Exercise No. 48

| | | |
|---|---|---|
| **1.** 22.566 | **4.** 71.083 | **7.** $11\frac{3}{10}$ |
| **2.** 12.996 | **5.** $1\frac{1}{2}$ | **8.** $17\frac{2}{5}$ |
| **3.** 25.637 | **6.** $15\frac{1}{4}$ | |

## Exercise No. 49

| | | | | |
|---|---|---|---|---|
| **1.** .32 | **3.** 8.3 | **5.** 22.0 | **7.** 4.042 | **9.** 4.376 |
| **2.** 1.6 | **4.** 11.8 | **6.** 2.93 | **8.** 5.792 | **10.** 15.852 |

## Exercise No. 50

| | | | |
|---|---|---|---|
| **1.** .048 | **6.** 1.61 | **11.** 53.3 | **16.** 549 |
| **2.** .782 | **7.** .0470 | **12.** .0036 | **17.** 450 |
| **3.** .1376 | **8.** 150.50 | **13.** .7802 | **18.** 89.25 |
| **4.** .676 | **9.** 1.305 | **14.** 2.85 | **19.** 3.744 |
| **5.** 8.379 | **10.** 18.02 | **15.** 2,115.52 | **20.** 6.913 |

## Exercise No. 51

| | | |
|---|---|---|
| **1.** .4 | **6.** 8030 | **11.** .51 |
| **2.** 537 | **7.** 1645 | **12.** 53,000 |
| **3.** 630 | **8.** 613.7 | **13.** 4 |
| **4.** 8521 | **9.** 31.416 | **14.** .03 |
| **5.** 3.7 | **10.** 8500 | **15.** $164.70 |

## Exercise No. 52

| | | |
|---|---|---|
| **1.** .087 | **6.** .078 | **11.** .532 |
| **2.** .0085 | **7.** .000063 | **12.** .00097 |
| **3.** 29.73 | **8.** .94 | **13.** $2.45 |
| **4.** 3.87 | **9.** $.25 | **14.** .5 |
| **5.** $.25 | **10.** .00387 | **15.** $1.25 |

## Exercise No. 53

| | | |
|---|---|---|
| **1.** 75.2 | **6.** .074 | **11.** 2.53 |
| **2.** .1267 | **7.** .03 | **12.** 25.3 |
| **3.** .525 | **8.** 2932 | **13.** 568.4 |
| **4.** 13 | **9.** .2821 | **14.** 28,700 |
| **5.** .4 | **10.** 2590 | **15.** .00039 |

## Exercise No. 54

| | |
|---|---|
| **1.** 1.4 | **6.** 155 |
| **2.** .021 | **7.** 27.12 |
| **3.** 350 | **8.** 57.3 |
| **4.** .02662 | **9.** 73.7 |
| **5.** 2.03 | **10.** 7.95 |

## Exercise No. 55

| | | | |
|---|---|---|---|
| **1.** 21.5 | **6.** 24.8 | **11.** 2.62 | **16.** 11.26 |
| **2.** 5.8 | **7.** 3.1 | **12.** 6.07 | **17.** 3.28 |
| **3.** 6.7 | **8.** 19.7 | **13.** 20.02 | **18.** 25.67 |
| **4.** 18.1 | **9.** 10.0 | **14.** 4.33 | **19.** 102.29 |
| **5.** 102.4 | **10.** .4 | **15.** 1.10 | **20.** 16.32 |

### Exercise No. 56

| | | | |
|---|---|---|---|
| **1.** 38.50 | **6.** 13.8 | **11.** 138.23 | **16.** 1.51 |
| **2.** 2.8 | **7.** 2.2 | **12.** 145.64 | **17.** 11.63 |
| **3.** 9.7 | **8.** 12.2 | **13.** 1.03 | **18.** 3.94 |
| **4.** 2.2 | **9.** 6.5 | **14.** 8.95 | **19.** .01 |
| **5.** 5.2 | **10.** 56.1 | **15.** 8.24 | **20.** 127.27 |

### Exercise No. 57

| (a) | (b) | (c) | | (d) |
|---|---|---|---|---|
| .125 | .0625 | .03125 | .015625 | .515625 |
| .25 | .1875 | .09375 | .046875 | .546875 |
| .375 | .3125 | .15625 | .078125 | .578125 |
| .5 | .4375 | .21875 | .109375 | .609375 |
| .625 | .5625 | .28125 | .140625 | .640625 |
| .75 | .6875 | .34375 | .171875 | .671875 |
| .875 | .8125 | .40625 | .203125 | .703125 |
| | .9375 | .46875 | .234375 | .734375 |
| | | .53125 | .265625 | .765625 |
| | | .59375 | .296875 | .796875 |
| | | .65625 | .328125 | .828125 |
| | | .71875 | .359375 | .859375 |
| | | .78125 | .390625 | .890625 |
| | | .84375 | .421875 | .921875 |
| | | .90625 | .453125 | .953125 |
| | | .96875 | .484375 | .984375 |

### Exercise No. 58

**1.** $30\overline{)395.40}$ = $13.18

**2.** $14.3 \times 9.4 = 134.42$ miles

**3.** $32 \times .087 = 2.784$ inches

**4.** $6.25\overline{)12.15}$ = 1.944 or $1.94 per yard

**5.** $358.4 \times 3.6 = 1290.24$

**6.** $5425.6 \times 5.8 = 31,468.48$ pounds

**7.** $2.4\overline{)8.9}$ = 3.7

**8.** $10.9\overline{)1000}$ = 92

**9.** $.0156\overline{)1.5}$ = 96

**10.** $20.8\overline{)457.6}$ = $22 \times \frac{1}{2}$ = 11 + 22 = 33

                                            Ans.

### Exercise No. 59

| | | |
|---|---|---|
| $\frac{1}{5}$ | $\frac{1}{4}$ | $\frac{1}{6}$ |
| $\frac{3}{10}$ | $\frac{3}{8}$ | $\frac{1}{3}$ |
| $\frac{2}{5}$ | $\frac{5}{8}$ | $\frac{2}{3}$ |
| $\frac{1}{2}$ | | |

### Exercise No. 60

| | | | | |
|---|---|---|---|---|
| **1.** .03 | **4.** .25 | **7.** 2. | **10.** .71 | **13.** .535 |
| **2.** 1. | **5.** .5 | **8.** .06 | **11.** .65 | **14.** .95 |
| **3.** .05 | **6.** .2 | **9.** .4 | **12.** .9 | **15.** 1.25 |

### Exercise No. 61

| | | |
|---|---|---|
| **1.** 20% | **6.** 25.3% | **11.** 23.8% |
| **2.** 6% | **7.** 253% | **12.** 45% |
| **3.** 10% | **8.** 12.5% | **13.** 60% |
| **4.** 40% | **9.** 100% | **14.** 1% |
| **5.** 47% | **10.** 2% | **15.** 50% |

### Exercise No. 62

| | | |
|---|---|---|
| **1.** 50% | **6.** $62\frac{1}{2}$% | **11.** $12\frac{1}{2}$% |
| **2.** 5% | **7.** $33\frac{1}{3}$% | **12.** $37\frac{1}{2}$% |
| **3.** 75% | **8.** $8\frac{1}{3}$% | **13.** $22\frac{2}{9}$% |
| **4.** $87\frac{1}{2}$% | **9.** 20% | **14.** $66\frac{2}{3}$% |
| **5.** 22% | **10.** 25% | **15.** $42\frac{6}{7}$% |

### Exercise No. 63

| | | |
|---|---|---|
| 50% | $\frac{1}{2}$ | .50 |
| $33\frac{1}{3}$% | $\frac{1}{3}$ | .333 |
| 25% | $\frac{1}{4}$ | .25 |
| $12\frac{1}{2}$% | $\frac{1}{8}$ | .125 |
| $37\frac{1}{2}$% | $\frac{3}{8}$ | .375 |
| $66\frac{2}{3}$% | $\frac{2}{3}$ | .666 |
| 75% | $\frac{3}{4}$ | .75 |
| $62\frac{1}{2}$% | $\frac{5}{8}$ | .625 |
| $87\frac{1}{2}$% | $\frac{7}{8}$ | .875 |

### Exercise No. 64

| | | |
|---|---|---|
| **1.** $\frac{1}{6}$ | **10.** .1 | **18.** 20% |
| **2.** 18% | **11.** $\frac{5}{8}$ | **19.** $\frac{5}{8}$ |
| **3.** $\frac{1}{3}$ | **12.** $\frac{1}{3}$ | **20.** $83\frac{1}{3}$ |
| **4.** $\frac{5}{8}$ | **13.** $\frac{7}{8}$ | **21.** $.83\frac{1}{3}$, $87\frac{1}{2}$%, .9 |
| **5.** 75% | **14.** $\frac{1}{3}$ | **22.** $\frac{5}{8}$, $66\frac{2}{3}$%, .675 |
| **6.** $62\frac{1}{2}$% | **15.** $\frac{1}{5}$ | **23.** $12\frac{1}{2}$%, $\frac{1}{5}$, .25 |
| **7.** $\frac{1}{7}$ | **16.** $\frac{1}{10}$ | **24.** $12\frac{1}{2}$%, $.16\frac{2}{3}$, $\frac{1}{5}$ |
| **8.** 19% | **17.** .07 | **25.** $\frac{2}{3}$, 75%, .8 |
| **9.** $\frac{7}{8}$ | | |

### Exercise No. 65

| | | |
|---|---|---|
| **1.** 7 | **6.** $1.60 | **11.** $72 |
| **2.** $6.10 | **7.** 540 | **12.** $45 |
| **3.** 208 | **8.** 110 | **13.** 81 |
| **4.** $30 | **9.** 483 | **14.** 150 |
| **5.** 93 | **10.** 140 | **15.** $1.25 |

### Exercise No. 66

| | | |
|---|---|---|
| **1.** .25 | **6.** 9.74 | **11.** $.10 |
| **2.** $.98 | **7.** 184.5 | **12.** $1.25 |
| **3.** $.05 | **8.** 8.5 | **13.** $.03 |
| **4.** $5.28 | **9.** $2.30 | **14.** $15.00 |
| **5.** 2.65 | **10.** $.23 | **15.** $.15 |

## Exercise No. 67

1. $.12 \times \$680 = \$81.60$
2. $\frac{3}{4}$ of $24 = 18$
3. $3\frac{1}{2}\%$ of $37,800 = .035 \times 37,800 = \$1323$
4. $\frac{4}{5}$ of $240 = 192$
5. $\frac{1}{5}$ of $\$35 = \$7$
6. $3\%$ of $\$42 = \$1.26$
7. $12\frac{1}{2}\% = \frac{1}{8}, \frac{1}{8} \times \$8720 = \$1090$
8. $10\%$ of $142.50 = 14.25, 14.25 + 142.50 = \$156.75$
9. $1\%$ of $\$350 = \$3.50$
10. $1\%$ of $\$760 = \$7.60, \frac{3}{4}$ of $\$7.60 = \$5.70$

## Exercise No. 68

1. 50%
2. $33\frac{1}{3}\%$
3. 25%
4. 28.6%
5. 150%
6. $66\frac{2}{3}\%$
7. 150%
8. 3%
9. $116\frac{2}{3}\%$
10. 3%

## Exercise No. 69

1. $\frac{120}{150} = \frac{4}{5} = 80\%$
2. $\frac{37\frac{1}{2}}{50} = \frac{6}{8} = 75\%$
3. $\frac{160}{250} = \frac{16}{25} = 64\%$
4. $\frac{450}{800} = \frac{45}{80} = \frac{9}{16} = 56\frac{1}{4}\%$
5. $\frac{12}{80} = \frac{3}{20} = 15\%$
6. $\frac{16}{25} = 64\%$
7. $\frac{6}{9} = \frac{2}{3} = 66\frac{2}{3}\%$
8. $\frac{25}{125} = \frac{1}{5} = 20\%$
9. $\frac{5}{40} = \frac{1}{8} = 12\frac{1}{2}\%$
10. $80 - 38 = 42, \frac{42}{80} = \frac{21}{40} = 52\frac{1}{2}\%$

## Exercise No. 70

1. 1500   3. $250   5. 300   7. 600   9. $20
2. 1000   4. 600   6. 1000   8. 6   10. 100

## Exercise No. 71

1. $\$15 \div \frac{3}{8} = \$40$
2. $\$21.30 \div \frac{1}{10} = \$213$
3. $\$9 \div \frac{3}{10} = \$30$
4. $\$50 \div \frac{20}{100} = \$250$
5. $\$27 \div \frac{3}{10} = 90$
6. $\$215 \div \frac{3}{4} = 286.666$ or $\$286.67$
7. $12 \div \frac{8}{100} = \$150$
8. $\$1200 \div \frac{1}{20} = \$24,000$
9. $\$84,000 \div \frac{4}{10} = \$210,000$
10. $\$19.20 \div 8 = \$2.40, \$2.40 \div \frac{1}{5} = \$12.00$

## Exercise No. 72

1. 14%
2. 54%
3. 69%
4. 96%
5. 92%
6. 14.3%
7. 76.8%
8. 82.3%
9. 56.4%
10. 61.5%
11. 19.38%
12. 65.72%
13. 23.5%
14. 134.68%
15. 80%

## Exercise No. 73

1. $16\frac{2}{3}\%$
2. 20%
3. 100%
4. 75%
5. 50%
6. 25%
7. 40%
8. 75%
9. $33\frac{1}{3}\%$
10. 50%

## Exercise No. 74

1. $\$495 - \$450 = \$45, \frac{45}{450} = 10\%$
2. $50,000 - 48,000 = 2000, \frac{2000}{50000} = 4\%$
3. $\$3.50 - \$3.00 = \$.50, \frac{50}{300} = \frac{1}{6} = 16\frac{2}{3}\%$
4. $456 - 380 = 76, \frac{76}{380} = 20\%$
5. $\frac{1}{8}$ of $3.80 = .475 = .48, 3.80 + .48 = \$4.28$
6. $\$50 \times 12 = \$600, \frac{600}{8000} = \frac{3}{40} = 7\frac{1}{2}\%$
7. Add $\$30 + \$15 + \$8 + \$7 = \$60$, $\frac{30}{60} = \frac{1}{2} = 50\%$
8. $\frac{75}{150} = 50\%, \frac{80}{175} = 45.71\%$ $50\% - 45.71\% = 4.29\%$
9. $950 - 725 = 225, \frac{225}{950} = \frac{9}{38} = 23.7\%$
10. $100\% - 15\% = 85\%, \$51 = 85\%$ of cost $\$51 \div \frac{85}{100} = \$51 \times \frac{20}{17} = \$60$ or $\$51 = 85\%, 1\% = \frac{51}{85}, \frac{51}{85} = \$.60$, $\$.60 \times 100 = \$60$

## Exercise No. 75

| Item | Rate of Discount | Discount | List Price | List Price |
|---|---|---|---|---|
| 1 | 37% | $140.00 | $379.00 | $239.00 |
| 2 | 21% | $156.00 | $749.00 | $593.00 |
| 3 | 14% | $5.00 | $36.95 | $31.95 |
| 4 | 35% | $7.00 | $19.95 | $12.95 |
| 5 | 41% | $7.00 | $16.95 | $9.95 |
| 6 | 24% | $12.00 | $49.98 | $37.98 |
| 7 | 1/3 off | $8.67 | $26.00 | $17.33 |
| 8 | 5% | $12.10 | $242.00 | $229.90 |
| 9 | 28% | $27.50 | $97.50 | $70.00 |
| 10 | 25% | $170.00 | $680.00 | $510.00 |

## Exercise No. 76

1. $2\%$ of $\$77 = \$1.54, \$77.00 - \$1.54 = \$75.46$ or $100\% - 2\% = 98\% \times \$77.00 = \$75.46$
2. $80\%$ of $\$12.50 = \frac{4}{5} \times \$12.50 = \$10.00$ or $\frac{1}{5}$ of $\$12.50 = \$2.50, \$12.50 - \$2.50 = \$10.00$
3. $66\%$ of $\$240$ or $\frac{2}{3} \times \$240 = \$160$
4. Discount is $\$1.25 - \$.75 = \$.50, \frac{50}{125} = \frac{2}{5} = 40\%$

### Exercise No. 76 (Continued)

5. $\frac{1}{5} \times 38.00 = 7.60$, $38.00 - 7.60 =$ $30.40
   or $100\% - 20\% = 80\%$, $80\%$ of $38.00 = $30.40

6. $10.98 - $5.00 = $5.98$, $\frac{598}{1098} = 55\%$
   $17.95 - $5.00 = $12.95$, $\frac{1295}{1795} = 72\%$

7. $45 - $19 = $26$, $\frac{26}{45} = 58\%$

8. $30\%$ of $49.75 = $14.92$, $49.75 - $14.92 = $34.83$
   $10\%$ of $34.83 = $3.48$, $34.83 + $3.48 = $38.31$ Ans. or
   $70\%$ of $49.75 = $34.83$, $34.83 + $3.48 = $38.31$ Ans.

9. $90\%$ of $29.95 = $26.95$

10. $90\%$ of $142.50 = 128.25$, $137.00 - $128.25 = $8.75$ (over)

### Exercise No. 77

1. $100 - 15\% = 85\%$, $85\%$ of what is $4,
   $4 \div \frac{85}{100} = $4 \times \frac{20}{17} = $4.70$

2. $100 - 20\% = 80\%$, $80\%$ of what is $22,
   $22 \div \frac{4}{5} = $22 \times \frac{5}{4} = $27.50$

3. $100 - 18\% = 82\%$, $82\%$ of what is $18,
   $18 \div \frac{82}{100} = $18 \times \frac{100}{82} = $21.95$

4. $100 - 25\% = 75\%$, $75\%$ of what is $10.50,
   $10.50 \div \frac{3}{4} = $10.50 \times \frac{4}{3} = $14$

5. $100 - 33\frac{1}{3}\% = 66\frac{2}{3}\%$, $66\frac{2}{3}\%$ of what is $22,
   $22 \div \frac{2}{3} = $22 \times \frac{3}{2} = $33$

### Exercise No. 78

1. $216
2. $289
3. $137.20
4. $418.95
5. $120

### Exercise No. 79

1. Total $3775, $\frac{3}{100}$ of $3775 = $113.25$
2. $125, $400 \times .025 = $3135$
3. $45 \times $95.50 = $4297.50$, $4297.50 \times .02 = $85.95$
4. $1830 \times .015 = 27.45$, $27.45 + 160 = $187.45$
5. $825 \times .12 = $99$, $750 \times .15 = $112.50$, $99 + $112.50 + $150 = $361.50$
6. $5\%$ of $100,000 = 5000$, $3\%$ of $115,500 = 3465$, $5000 + 3465 = $8465$
7. $\frac{72}{1800} = \frac{12}{300} = \frac{1}{25} = 4\%$
8. $\frac{14}{70} = \frac{1}{5} = 20\%$

### Exercise No. 80A

a. $204.60 \div $.66 = 310$ boxes
b. $.44 \times 310 = $136.40$
c. $204.60 - $13.64 - $136.40 = $54.56$
d. $66¢ - 44¢ = 22¢$
e. $\frac{22¢}{44¢} = \frac{1}{2} = 50\%$
f. $\frac{22¢}{66¢} = \frac{1}{3} = 33\frac{1}{3}\%$
g. $\frac{$204.60 - $13.64 - $136.40}{$.44 \times 310 + $13.64} = \frac{$54.56}{$150.04}$
   $= 36\%$
h. $\frac{$54.56}{$204.60} = 27\%$

### Exercise No. 80B

1. Selling Price
2. First Cost
3. Gross Cost
4. Overhead
5. Gross Cost
6. Overhead
7. Selling Price
8. Loss
9. Gross Cost
10. Overhead

### Exercise No. 81

| | Cost | Gross Profit | % Profit on Cost | % Profit on Selling Price | Selling Price |
|---|---|---|---|---|---|
| 1. | $75 | $25 | $33\frac{1}{3}\%$ | 25% | $100 |
| 2. | $400 | $100 | 25% | 20% | $500 |
| 3. | $200 | $100 | 50% | $33\frac{1}{3}\%$ | $300 |
| 4. | $160 | $40 | 25% | 20% | $200 |
| 5. | $1450 | $550 | 37.9% | $27\frac{1}{2}\%$ | $2000 |

### Exercise No. 82

1.

| | First Cost | Selling Price | Overhead in % of Cost | Overhead in Dollars | Markup in % of Cost | Markup in Dollars | Profit in Dollars |
|---|---|---|---|---|---|---|---|
| A | $300 | $500 | 50% | $150 | $66\frac{2}{3}\%$ | $200 | $50 |
| B | $150 | $270 | 40% | $60 | 80% | $120 | $60 |
| C | $250 | $475 | 40% | $100 | 90% | $225 | $125 |
| D | $150 | $290 | 54.6% | $82 | 93% | $140 | $58 |
| E | $750 | $1000 | 20% | $150 | $33\frac{1}{3}\%$ | $250 | $100 |

| | Profit in % of Sales | Profit in % of Cost |
|---|---|---|
| A | 10% | $16\frac{2}{3}\%$ |
| B | 22.2% | 40% |
| C | 26.3% | 50% |
| D | 20% | 38.6% |
| E | 10% | 13.3% |

2. S.P. $= 100 + 35 + 55 = $190.00$ Rate of Profit $= \frac{55}{190} = 28.9\%$

3. S.P. $=$ Cost $+$ Overhead $+$ Profit, $\frac{3}{10}$ of $8 = $2.40$, $\frac{2}{10}$ of $8 = $1.60$
   or $8 + $4 = $12$. S.P., Profit $= 20\%$ of $8 = $1.60$

4. $20\% + 45\% = 65\%$, $\frac{65}{100} \times 250 = $162.50$
   $162.50 + $250 = $412.50$

## Exercise No. 82 (Continued)

5. S.P. = Cost + Overhead + Profit
$33\frac{1}{3}\%$ + 10% = $43\frac{1}{3}\%$, $43\frac{1}{3}\%$ of $6 = $2.60,
$2.60 + $6 = $8.60 ÷ 24 = $.36 per can

6. Cost + Overhead + Profit = S.P.
235 + 90 + ? = 475, profit = 150, $\frac{150}{475}$ = 32%

7. $\frac{30}{100}$ = 30%, $\frac{45}{175}$ = $\frac{9}{35}$ = 26%

8. $\frac{1}{5}$ = 20%

9. 60% of $140 = $84

10. 25 × $.88 = $22 − $8 = $14, $\frac{14}{22}$ = $.636
= 63.6%

11. 25% of 2000 = 500, $\frac{3}{8}$ of $2000 = $750,
$750 − $500 = $250

## Exercise No. 83

| | I | A | | | I | A |
|---|---|---|---|---|---|---|
| 1. | $11.50 | $586.50 | | 16. | $18.00 | $1218.00 |
| 2. | $10.00 | $210.00 | | 17. | $34.00 | $834.00 |
| 3. | $10.50 | $360.50 | | 18. | $16.50 | $316.50 |
| 4. | $33.00 | $583.00 | | 19. | $22.50 | $1022.50 |
| 5. | $16.00 | $416.00 | | 20. | $7.50 | $507.50 |
| 6. | $51.00 | $901.00 | | 21. | $20.00 | $520.00 |
| 7. | $36.00 | $1236.00 | | 22. | $6.25 | $256.25 |
| 8. | $45.00 | $945.00 | | 23. | $40.00 | $440.00 |
| 9. | $15.60 | $405.60 | | 24. | $6.00 | $606.00 |
| 10. | $30.00 | $1530.00 | | 25. | $12.00 | $412.00 |
| 11. | $27.00 | $627.00 | | 26. | $26.00 | $326.00 |
| 12. | $11.25 | $461.25 | | 27. | $3.50 | $178.50 |
| 13. | $49.50 | $949.50 | | 28. | $7.50 | $1507.50 |
| 14. | $12.95 | $382.95 | | 29. | $3.00 | $203.00 |
| 15. | $24.75 | $574.75 | | 30. | $40.00 | $840.00 |

## Exercise No. 84

1. 2%
2. 2 yrs.
3. $2\frac{1}{2}\%$
4. $1\frac{1}{2}$ yrs.
5. 4%
6. 8 months
7. 5%
8. $2\frac{1}{2}$ yrs.
9. 6%
10. 3 yrs.

## Exercise No. 85

1. $3.40
2. $8.65
3. $14.50
4. $.30
5. $9.21
6. $1.11
7. $.92
8. $6.43
9. $.01
10. $42.59

## Exercise No. 86

| 1. | $2.50 | 6. | $2.10 | 11. | $2.88 | 16. | $3.14 |
|---|---|---|---|---|---|---|---|
| 2. | $15.00 | 7. | $25.00 | 12. | $11.55 | 17. | $4.32 |
| 3. | $9.00 | 8. | $33.00 | 13. | $2.67 | 18. | $5.88 |
| 4. | $2.00 | 9. | $6.75 | 14. | $9.33 | 19. | $12.88 |
| 5. | $10.00 | 10. | $1.05 | 15. | $9.80 | 20. | $.98 |

## Exercise No. 87

1. $106.09
2. $270.60
3. $522.84
4. $848.96
5. $761.50
6. $61.73
7. $228.94
8. $109.66
9. $467.34
10. $3611.22

## Exercise No. 88

1. $200 × 1.8
166 = $237.32
2. 6.5%
3. (a) $1104.49  (b) $1282.04  (c) $2107.18
4. 1.2689 × $400 = $507.56,
507.56 ÷ 2 = 253.73
5. 1.72677 × $2500 = $4316.93

## Exercise No. 89

| | Discount | Net Proceeds |
|---|---|---|
| 1. | $2.70 | $537.30 |
| 2. | $3.50 | $346.50 |
| 3. | $2.75 | $217.25 |
| 4. | $1.00 | $199.00 |
| 5. | $4.50 | $145.50 |

## Exercise No. 90

1. The actual interest is 3% × 12 = 36%

To calculate the number of months and amount of the last payment we make the following table:

| Month | Amount of Loan | Payment | Interest | Payment on Principal | New Balance |
|---|---|---|---|---|---|
| 1 | $80.00 | $10 | $2.40 | $7.60 | $72.40 |
| 2 | $72.40 | $10 | $2.17 | $7.83 | $64.57 |
| 3 | $64.57 | $10 | $1.94 | $8.06 | $56.51 |
| 4 | $56.51 | $10 | $1.70 | $8.30 | $48.21 |
| 5 | $48.21 | $10 | $1.45 | $8.55 | $39.66 |
| 6 | $39.66 | $10 | $1.19 | $8.81 | $30.85 |
| 7 | $30.85 | $10 | $ .93 | $9.07 | $21.78 |
| 8 | $21.78 | $10 | $ .65 | $9.35 | $12.43 |
| 9 | $12.43 | $10 | $ .37 | $9.63 | $ 2.80 |
| 10 | $ 2.80 | $ 2.88 | $ .08 | $2.80 | $ 0 |

From the table we see that the last payment is $2.88 and that it takes 10 months to repay the loan.

2. To find the total amount paid and the last payment, we need to calculate the following table:

| Month | Amount of Loan | Payment | Interest | Payment on Principal | New Balance |
|---|---|---|---|---|---|
| 1 | $200.00 | $30 | $4.00 | $26.00 | $174.00 |
| 2 | $174.00 | $30 | $3.48 | $26.52 | $147.48 |
| 3 | $147.48 | $30 | $2.95 | $27.05 | $120.43 |
| 4 | $120.43 | $30 | $2.41 | $27.59 | $ 92.84 |
| 5 | $ 92.84 | $30 | $1.88 | $28.12 | $ 64.70 |
| 6 | $ 64.70 | $30 | $1.29 | $28.71 | $ 35.99 |
| 7 | $ 35.99 | $30 | $ .72 | $29.28 | $ 6.71 |
| 8 | $ 6.71 | $ 6.84 | $ .35 | $ 6.73 | $ 0 |

### Exercise No. 90 *(Continued)*

So the total amount is $(7 \times \$30) + \$6.84 = \$216.84$. We could also get the total amount paid by adding the sums of the INTEREST and PAYMENT ON PRINCIPAL columns.

3.

|  | MONTH | BALANCE |
|---|---|---|
| 2000 × .088 = 176 FINANCE CHARGE | 1 | 2176.00 |
| 2000 + 176 = 2176 AMOUNT FINANCED | 2 | 1994.67 |
| 2176 ÷ 12 = 181.33 | 3 | 1813.34 |
| NOW MAKE TABLE AT THE RIGHT BY | 4 | 1632.01 |
| SUBTRACTING SUCCESSIVELY THE | 5 | 1450.68 |
| PAYMENT AMOUNT OF 181.33 | 6 | 1269.35 |
|  | 7 | 1088.02 |
| TOTAL THE MONTHLY BALANCES | 8 | 906.69 |
| TO GET 14,144.22 AND DIVIDE | 9 | 725.36 |
| BY 12. 14,144.22 ÷ 12 = 1178.69 | 10 | 544.03 |
| THIS IS THE AVERAGE LOAN | 11 | 362.70 |
|  | 12 | 181.37 |
| 176 ÷ 1178.69 = .149 | + |  |
| OR 15 PERCENT |  | 14,144.22 |

### Exercise No. 91

| | | | |
|---|---|---|---|
| **1.** 12 | **10.** 4 | **19.** 3 | **28.** 2000 |
| **2.** 16 | **11.** 480 | **20.** 16 | **29.** 24 |
| **3.** 8 | **12.** $5\frac{1}{2}$ | **21.** 4 | **30.** 4 |
| **4.** 2240 | **13.** 60 | **22.** 500 | **31.** 365 |
| **5.** $2\frac{5}{8}$ | **14.** 7 | **23.** 320 | **32.** $16\frac{1}{2}$ |
| **6.** 2 | **15.** 366 | **24.** 144 | **33.** 100 |
| **7.** 12 | **16.** 1760 | **25.** 60 | **34.** 32 |
| **8.** $31\frac{1}{2}$ | **17.** 20 | **26.** 36 | **35.** 5280 |
| **9.** 24 | **18.** 12 | **27.** 7 | **36.** 10 |

### Exercise No. 92

**1.** 8 ft. 2 in.  
**2.** 28 lb. 2 oz.  
**3.** 17 qt.  
**4.** 7 hr. 5 min.  
**5.** 3 yd. 2 ft. 2 in.  
**6.** 11 bu. 3 pk.  
**7.** 7 lb. 15 oz.  
**8.** 5 pt. 4 oz.  
**9.** 19 ft. 7 in.  
**10.** 16 hr. 13 min. 25 sec.

### Exercise No. 93

**1.** 5 weeks 6 days  
**2.** 3 yr. 10 mo.  
**3.** 4 ft. 9 in.  
**4.** 1 min. 28 sec.  
**5.** 4 tons 1200 lb.  
**6.** 1 mi. 880 yd.  
**7.** 1 lb. 10 oz.  
**8.** 5 yd. 2 ft.  
**9.** 2 gal. 3 qt.  
**10.** 3 lb. 11 oz.

### Exercise No. 94

**1.** 7 hr. 45 min.  
**2.** 11 gal.  
**3.** 7 lb. 2 oz.  
**4.** 32 yd. 2 ft.  
**5.** 25 qt. 1 pt.  
**6.** 5 tons 1600 lb.  
**7.** 7 mi. 880 yd.  
**8.** 7 hr. 15 min.  
**9.** 32 bu. 2 pk.  
**10.** 7 qt. 28 fl. oz.

### Exercise No. 95

**1.** 1 hr. 33 min.  
**2.** 2 yd. 20 in.  
**3.** 3 ft. $9\frac{2}{3}$ in.  
**4.** 2 pt. $13\frac{1}{2}$ fl. oz.  
**5.** 2 bu. 1 pk.  
**6.** 8 min. 30 sec.  
**7.** 1 qt. $1\frac{1}{8}$ pt.  
**8.** 1 qt. $14\frac{2}{3}$ fl. oz.  
**9.** 2 lb. $12\frac{1}{2}$ oz.  
**10.** 31 in.

### Exercise No. 96

**1.** $11{:}50 - 7{:}30 = 4$ hr. 20 min. $4\frac{1}{3} \times \$2.75 = 11.92$

**2.** 3 lb. 12 oz. + 4 lb. 10 oz. $= 8\frac{3}{8} \times \$.30 = \$2.51$

**3.** 5 gal. $-$ 1 gal. 3 qt. = 3 gal. 1 qt.

**4.** 2 hr. 20 min. + 1 hr. 45 min. + 35 min. = 4 hr. 40 min.

**5.** 4:25 P.M.

**6.** 12 lb. 8 oz. ÷ 5 = 2 lb. 8 oz.

**7.** 10 yd. 8 in. ÷ 4 = 2 yd. 20 in.

**8.** 8 lb. 10 oz. × 8 = 69 lb.

**9.** 2 pk. 2 qt. ÷ 3, 2 pk. = 16 qt. + 2 qt. = 18 qt. ÷ 3 = 6 qt.

**10.** 3 hr. 15 min. × 24 = 78 hr. ÷ 8 = $9\frac{3}{4}$ work days

### Exercise No. 97

| | |
|---|---|
| **1.** 32 in. | **16.** 2 qt. |
| **2.** 180 fl. oz. | **17.** 132 ft. |
| **3.** 1 hr. 25 min. | **18.** $1\frac{1}{4}$ ft. |
| **4.** 3 gal. 5 pt. | **19.** $6\frac{2}{3}$ yd. |
| **5.** $2\frac{1}{2}$ lb. | **20.** $2\frac{2}{3}$ yd. |
| **6.** 25 pt. | **21.** $2\frac{1}{8}$ lb. |
| **7.** 13 ft. | **22.** $1\frac{5}{6}$ hr. |
| **8.** 3 bu. 2 pk. | **23.** 2 oz. |
| **9.** 24 in. | **24.** $\frac{1}{4}$ ton |
| **10.** 12 min. | **25.** 528 ft. |
| **11.** $6\frac{1}{2}$ qt. | **26.** 1 qt. 16 fl. oz. |
| **12.** 14 oz. | **27.** $\frac{2}{3}$ yd. |
| **13.** $3\frac{1}{2}$ pt. | **28.** 4 ft. |
| **14.** $4\frac{3}{4}$ bu. | **29.** 40 qt. |
| **15.** $\frac{5}{8}$ lb. | **30.** 8 fl. oz. |

### Exercise No. 98

| | | |
|---|---|---|
| **1.** 2.54 | **9.** 250 mm | **16.** 5.823 M |
| **2.** 3.94 | **10.** 1200 cm | **17.** .005 Km |
| **3.** 30 | **11.** 3410 M | **18.** 800 M |
| **4.** 25.4 | **12.** 7 mm | **19.** 90 M |
| **5.** $\frac{3}{10}$ | **13.** 2.57 M | **20.** 4 M |
| **6.** 2300 M | **14.** 582.3 cm | **21.** 109.36 |
| **7.** 400 cm | **15.** 5.823 M | **22.** 437.44 |
| **8.** 4.825 Km. | | |

**Exercise No. 98** *(Continued)*

**23.** The mile run—by 120 yd.
**24.** 66 in. × 2.54 cm = 167.64 cm
**25.** 35 mm. = 3.5 cm   3.5 ÷ 2.54 = 1.37 in.
**26.** 336 × $\frac{5}{8}$ = 210 miles (approximately)
**27.** Divide by $\frac{5}{8}$   50 × $\frac{8}{5}$ = 80 Km per hr. (approximately)

## Exercise No. 99

**1.** 140 ÷ 2.2 = 63.6 kg.
**2.** **(a)** 500 gm. butter     **(c)** 25 gm. cinnamon
  **(b)** 1 kg. apples     **(d)** 2 kg. potatoes
**3.** 14.2 gm.
**4.** 2 oz. = 56.8 gm. $.2\overline{)56.8}$ = $2\overline{)568}$ = 284 pills
**5.** 25 × 2.2 = 55 lb.

## Exercise No. 100

**1.** **(a)** 5:00 P.M.   **(c)** 3:00 P.M.   **(e)** 4:00 P.M.
  **(b)** 6:00 P.M.   **(d)** 3:00 P.M.
**2.** Set it back 3 hr.
**3.** 30°
**4.** 7:30 P.M.
**5.** 5:00 P.M.
**6.** 9:30 Chicago is 7:30 San Francisco, 2:45 P.M.
**7.** Same zone—2:45 P.M.
**8.** 9:30 Los Angeles is 10:30 Denver 12:50 P.M.
**9.** 11:15 New Orleans is 12:15 New York, 5:55 P.M.
**10.** Same zone—4:40 P.M.

## Exercise No. 101

**1.** **(a)** 0035   **(c)** 0300   **(e)** 1135
  **(b)** 1235   **(d)** 2028
**2.** **(a)** 12:45 A.M.   **(c)** 12:03 P.M.   **(e)** 4:35 A.M.
  **(b)** 5:55 P.M.   **(d)** 7:50 P.M.
**3.** 11 hr. 50 min.
**4.** 2215 E.S.T.

## Exercise No. 102

**1.** 90°
**2.** 45°
**3.** 180°
**4.** 12:20
**5.** **(a)** acute   **(c)** right   **(e)** obtuse
  **(b)** acute   **(d)** straight
**6.** 15°
**7.** 35°
**8.** 80°
**9.** 50°
**10.** 105°

## Exercise No. 103

**1.** 76°   scalene   acute
**2.** 30°   scalene   right
**3.** 90°   isosceles   right
**4.** 30°   scalene   obtuse
**5.** 42°   scalene   acute
**6.** 50°   isosceles   acute
**7.** 60°   equilateral   acute-equiangular
**8.** 116°   scalene   obtuse

## Exercise No. 104

**1.** 160 ft.   **5.** 62 ft.   **8.** $58\frac{1}{2}$ ft.
**2.** 18 ft.   **6.** 72 ft.   **9.** 4 ft.
**3.** 25 ft.   **7.** 1840 ft.   **10.** 80 yards.
**4.** 540 ft.

## Exercise No. 105

**1.** **(a)** 1 in.   **(b)** $1\frac{1}{2}$ in.   **(c)** $1\frac{3}{4}$ in.
**2.** $2\frac{1}{4}$ in.
**3.** $1\frac{1}{4}$ in.

## Exercise No. 106

**1.** 3.14 × 5 = 15.7 ft.
**2.** 3.14 × 10 = 31.4 in.
**3.** $\frac{22}{7} \times \frac{15}{2} = 23\frac{4}{7}$ in.
**4.** $\frac{22}{7} \times \frac{28}{1}$ = 88 ft.
**5.** $\frac{22}{7} \times 2\frac{1}{6} = \frac{22}{7} \times \frac{13}{6} = \frac{143}{21}$ = 6.8 ft.
  5280 ÷ 6.8 = 776.4 turns
**6.** 3.14 × 8000 = 25,120 miles
**7.** $\frac{22}{7}$ × 840 = 2640, 5280 ÷ 2640 = 2 (times)

## Exercise No. 107

**1.** 144 sq. in.   **6.** $30\frac{1}{4}$ sq. yd.
**2.** 9 sq. ft.   **7.** $272\frac{1}{4}$ sq. ft.
**3.** 1296 sq. in.   **8.** 43,560 sq. ft.
**4.** 160 sq. rd.   **9.** 1 sq. yd.
**5.** 640 acres   **10.** 1 sq. ft.

## Exercise No. 108

**1.** 21 sq. ft.   **6.** 1 sq. ft.
**2.** $\frac{1}{6}$ sq. ft.   **7.** 6 sq. yd.
**3.** 600 sq. yd.   **8.** 121 sq. ft.
**4.** $20\frac{4}{25}$ acres   **9.** $2\frac{1}{4}$ sq. ft.
**5.** $\frac{1}{2}$ sq. ft.   **10.** $28\frac{4}{9}$ sq. yd.

## Exercise No. 109

**1.** 9 sq. ft.   **4.** 24 sq. yd.
**2.** $3\frac{1}{8}$ sq. ft.   **5.** 216 sq. ft.
**3.** 45 sq. ft.

## Exercise No. 110

| | |
|---|---|
| **1.** 120 sq. in. | **5.** 59.5 sq. in. |
| **2.** 72 sq. in. | **6.** 3 sq. ft. |
| **3.** 48 sq. ft. | **7.** 2475 sq. ft. |
| **4.** 81 sq. in. | **8.** 2500 sq. ft. |

## Exercise No. 111

| | |
|---|---|
| **1.** 75.46 sq. in. | **5.** 3850 sq. ft. |
| **2.** 24.64 sq. in. | **6.** $1018\frac{2}{7}$ sq. in. |
| **3.** $1257\frac{1}{7}$ sq. in. | **7.** 616 sq. in. |
| **4.** 1386 sq. in. | **8.** 1386 sq. ft. |

## Exercise No. 112

**1.** $160 \times 160 = 25,600$ sq. ft., $5 \times 8 = 40$ sq. ft., $25,600 \div 40 = 640$ Ans.

**2.** $\frac{22}{7} \times 12 \times 12 = 452\frac{4}{7}$ sq. in.

**3.** $18 \times 12 = 216$ sq. ft., area of tile $\frac{3}{4} \times \frac{3}{4} = \frac{9}{16}$ sq. ft., $216 \div \frac{9}{16}$ or $216 \times \frac{16}{9} = 384$ tiles

**4.** $A = \frac{1}{2}bh = \frac{1}{2}(12 \times 18) = 108$ sq. in. $\times 10 = 1080$ sq. in. $\div 144 = 7.5$ sq. ft.

**5.** $120 \times 400 = 48,000$ sq. ft. $- 43,560$ (acre) $= 4,440$ sq. ft. more

**6.** $A = \frac{1}{2}bh = 6 \times 8 = 48$ sq. in. $200 \times 144 = 28,800; 28,800 \div 48 = 600$

**7.** $A = \frac{1}{2}bh = 4\frac{1}{2} \times 15 = \frac{135}{2} = 67\frac{1}{2}$ sq. ft.

**8.** $A = \pi r^2$, $A$ (pool) $= \frac{22}{7} \times (\frac{15}{2})^2 = 176\frac{11}{14}$, $A$ pool & walk $= \frac{22}{7} \times (\frac{21}{2})^2 = 346\frac{1}{2}$, $346\frac{7}{14} - 176\frac{11}{14} = 169\frac{5}{7} = 169\frac{5}{7}$ area of walk

**9.** $\frac{22}{7} \times (\frac{3}{2})^2 = 7\frac{1}{14}$, $\frac{22}{7} \times (3)^2 = 28\frac{4}{14}$, $28\frac{4}{14} - 7\frac{1}{14} = 21\frac{3}{14}$

**10.** $\frac{22}{7} \times (5)^2 = 78\frac{4}{7} \times \frac{15}{100} = \$11.78$

## Exercise No. 113

**1.** $V = lwh$   $18 \times 12 \times 8 = 1728$ cu. in.

**2.** $V = lwh$   $10 \times 6\frac{1}{2} \times 4 = 260$ cu. ft.

**3.** $3 \times 3 \times \frac{2}{3} = 6$ cu. ft.

**4.** $3 \times 3 \times 8 = 72$ cu. ft., $72 \div 36 = 2$ lb. Ans.

**5.** $2\frac{2}{3} \times 3 \times 2\frac{1}{4} = \frac{8}{3} \times \frac{3}{1} \times \frac{9}{4} = 18$ cu. ft.

**6.** 8 ft. $\times$ 50 ft. $\times \frac{1}{2}$ ft. $= 200$ cu. ft.; $200 \div 27 = 7.4$ yd.; $7.4 \times \$20 = \$148$

**7.** $6 \times 4 \times 2 = 48$, $48 \div \frac{1}{2} = 96$ Ans.

## Exercise No. 114

| | |
|---|---|
| **1.** 64 cu. in. | **6.** $75\frac{3}{7}$ cu. in. |
| **2.** 63 cu. in. | **7.** 504 cu. in. |
| **3.** 360 cu. in. | **8.** 144 cu. in. |
| **4.** 480 cu. in. | **9.** 8 cu. ft. |
| **5.** 384 cu. in. | **10.** 231 cu. ft. |

## Exercise No. 114 *(Continued)*

**11.** $V = \pi r^2 h \, \frac{22}{7} \times (100)^2 \times 91 = 2,860,000$ cu. ft.

**12.** $V = \frac{22}{7} \times 30 \times 30 \times 40 = 113,143 \times 7.5 = 848,572$ gal.

**13.** $\frac{22}{7} \times \frac{5}{2} \times \frac{5}{2} \times 20 = 393 \times 67 = 26,331$ lb.

**14.** $V = Ah$ where $A = \frac{1}{2}bh = \frac{1}{2}(4 \times 5) \times 90 = 900 \times 7\frac{1}{2} = 6750$ gal.

## Exercise No. 115

| | |
|---|---|
| **1.** $\frac{2}{12}$ or $\frac{1}{6}$ | **11.** 5 : 10 or 1 : 2 |
| **2.** $\frac{3}{60}$ or $\frac{1}{20}$ | **12.** 2 : 32 or 1 : 16 |
| **3.** $\frac{4}{10}$ or $\frac{2}{5}$ | **13.** 20 : 80 or 1 : 4 |
| **4.** $\frac{1}{5}$ | **14.** 400 : 50 or 8 : 1 |
| **5.** $\frac{3}{8}$ | **15.** 10 : 60 or 1 : 6 |
| **6.** $\frac{15}{5}$ or $\frac{3}{1}$ | **16.** 48 : 12 or 4 : 1 |
| **7.** $\frac{24}{12}$ or $\frac{2}{1}$ | **17.** 3 : 21 or 1 : 7 |
| **8.** $\frac{72}{108}$ or $\frac{2}{3}$ | **18.** 50 : 20 or 5 : 2 |
| **9.** $\frac{5}{\frac{1}{2}}$ or $\frac{10}{1}$ | **19.** 45 : 90 or 1 : 2 |
| **10.** $\frac{8\frac{1}{2}}{\frac{1}{2}}$ or $\frac{17}{1}$ | **20.** 2 : 8 or 1 : 4 |

## Exercise No. 116

**1.** 12 : 34 or 6 : 17

**2.** (a) 10 : 40 or 1 : 4    (b) 10 : 30 or 1 : 3

**3.** 80 : 100 or 4 : 5

**4.** 4 : 20 or 1 : 5

**5.** 5280 ft. $\div$ 1 mile, 52,800 ft. $= 10$ miles, $52,800 \times 12$ in. $= 633,600$ inches in 10 miles, 1 : 633,600 Ans.

## Exercise No. 117

| | |
|---|---|
| **1.** 8 | **4.** 7 |
| **2.** 6 | **5.** $2\frac{1}{4}$ |
| **3.** 24 | |

**6.** $6 : 26 :: 18 : ?$, $18 \times 26 = 468$, $468 \div 6 = 78\cent$ Ans.

**7.** $1 : 2 :: 2 : ?$, $2 \times 2 = 4$, $4 \div 1 = 4$ ans.

**8.** $2 : 3 :: ? : 36$, $2 \times 36 = 72$, $72 \div 3 = 24$ Ans.

**9.** $5 : 80 :: ? : 200$, $5 \times 200 = 1000$, $1000 \div 80 = 12\frac{1}{2}$ Ans.

**10.** $4 : 320 :: ? : 400$, $4 \times 400 = 1600$, $1600 \div 320 = 5$ Ans.

## Exercise No. 117B

| | |
|---|---|
| **1.** 60 mph | **4.** 5 mph |
| **2.** 30 mph | **5.** 15 mph |
| **3.** 13 mph | |

### Exercise No. 118

**1.** Age 13    **3.** Age 13 to 14
**2.** Age 8    **4.** No, 12 to 13
**5.** $95 - 50 = 45$, $45 \div 6 = 7\frac{1}{2}$ lb. per yr. average
**6.** $113 - 55 = 58$, $58 \div 6 = 9\frac{2}{3}$ lb., $9\frac{2}{3}$ lb. $- 7\frac{1}{2} = 2\frac{1}{6}$ lb.
**7.** 11 to 14
**8.** $9\frac{1}{3}$ compared to 12

### Exercise No. 118B

**1.** Food and rent
**2.** 15%
**3.** $60
**4.** Food $360
**5.** 2 : 5

### Exercise No. 119

**1.** 144°
**2.** 22°
**3.** Food and Miscellaneous
**4.** $\frac{3}{10}$
**5.** $\frac{7}{50}$

### Exercise No. 120

**1.** $+5$    **6.** $-32°$
**2.** $+15$    **7.** $-1000$ ft.
**3.** $-10$    **8.** 1858
**4.** $-5\%$    **9.** $-55°$
**5.** $+\$50$    **10.** $-18°$

### Exercise No. 121

**1.** $+8$    **6.** $+7$    **11.** $+11$
**2.** $+3$    **7.** $+6$    **12.** $+4$
**3.** $-7$    **8.** $-11$    **13.** $+6$
**4.** $-8$    **9.** $-5$    **14.** $-7$
**5.** $+3$    **10.** $-16$    **15.** $-9$

### Exercise No. 122

**1.** 6    **6.** 13
**2.** 11    **7.** 14
**3.** $-14$    **8.** 99
**4.** 1    **9.** 3
**5.** $-4$    **10.** $-18$

### Exercise No. 123

**1.** 12    **9.** 13
**2.** $-33$    **10.** 22
**3.** $-25$    **11.** $-36$
**4.** $-15a$    **12.** $-27$
**5.** $-7c$    **13.** 30
**6.** $2b$    **14.** $3a$
**7.** $15a - 2b$    **15.** $-25b$
**8.** $11a - 8b$    **16.** $-2c$

# INDEX